Shape Memory Materials

Shape Memory Materials

By

Arun D I

Chakravarthy P

Arockiakumar R

Santhosh B

CRC Press
Taylor & Francis Group
Boca Raton London New York

CRC Press is an imprint of the
Taylor & Francis Group, an **informa** business

CRC Press
Taylor & Francis Group
6000 Broken Sound Parkway NW, Suite 300
Boca Raton, FL 33487-2742

Printed and bound by CPI Group (UK) Ltd, Croydon, CR0 4YY on acid-free paper

International Standard Book Number-13: 978-0-8153-5969-2 (Hardback)

Library of Congress Cataloging-in-Publication Data

Names: I., Arun D., author. | P., Chakravarthy, author. | Kumar, Arockia, author. | B., Santhosh, author.
Title: Shape memory materials / Arun D I, Chakravarthy P, Arockia Kumar and Santhosh B.
Description: First edition. | Boca Raton, FL : CRC Press/Taylor & Francis Group, 2018. | "A CRC title, part of the Taylor & Francis imprint, a member of the Taylor & Francis Group, the academic division of T&F Informa plc." | Includes bibliographical references and index.
Identifiers: LCCN 2017059462| ISBN 9780815359692 (hardback : acid-free paper) | ISBN 9781351119948 (ebook)
Subjects: LCSH: Shape memory alloys. | Shape memory polymers.
Classification: LCC TA487 .I323 2018 | DDC 620.1/125--dc23
LC record available at https://lccn.loc.gov/2017059462

Visit the Taylor & Francis Web site at
http://www.taylorandfrancis.com

and the CRC Press Web site at
http://www.crcpress.com

Contents

Foreword

This book introduces a wide range of shape memory materials, such as alloys, ceramics, gels, polymers, and composites, but focuses mainly on polymer materials. Readers can study physical properties related to shape recovery, the mechanism of shape memory effect, typical materials, the history of material development for each shape memory material, and differences among shape memory materials. This book is suitable for beginners who start to learn about shape memory materials and for experts to re-realize and come up with new concepts for designing new materials.

Dr. Yoko Yamabe-Mitarai
Research Center for Structural Materials, Deputy Director
Energy Infrastructure Materials Field, Field Coordinator
High Temperature Materials Design Group, Group Leader
NIMS
Sengen, Japan

Preface

Additive manufacturing or 3D printing is one of the latest trends in material science and is even evolving into multidimensional printing beyond three dimensions through everyday innovations and research. When 3D-printed materials change shape with time in response to specific stimuli, this is known as *4D printing*. Universities and research organizations across the world have heavily invested their intellectual resources into this evolving trend in material science. Multidimensional printing can be realized with the use of shape-morphing and shape memory materials in place of conventional 3D-printed raw materials. Shape-morphing and shape memory materials thus form the basis of such innovations. The origins of shape memory materials can be traced back to the 1940s, when Ti-Ni alloys were explored for their memorizing abilities. Research on Ti-Ni-based materials reigned in the arena of shape memory smart materials until gels, ceramics, and polymers evolved as alternative options. The advantages of newer technologies over old existing techniques become the impetus for all innovations. Thus, studies on shape memory polymers over alloys have taken the lead.

Many peer-reviewed articles on shape memory materials have been published, covering studies on the memorizing behavior of alloys and polymers. This work adopts a unique narrative style by defining smart materials and taking the reader on a journey through the various shape memory materials and their corresponding activation techniques, characterizations, response stimuli, syntheses, and applications. Shape memory materials suit the demands of aerospace applications, and industry-specific studies have been initiated on the remote activation of these materials by indirect thermal or electrical actuation. Electroactivity demands studies on conductivity; these have been reviewed and reported chronologically in this work. The prospects of shape memory materials are discussed, befitting smart actuations for aerospace and other applications.

Graduate students working in materials science and interested faculties, industries, and innovators can make use of this book to understand the recent advances in the field of shape memory materials. This book comprehensively reviews the evolution of smart materials through the lessons learned from nature to the current scenario.

The first chapter gives an overview of the topic, with illustrations from nature that inspired researchers to develop newer materials with improved properties. Chapters 2 through 6 deal with various shape memory materials and their advantages and disadvantages, illustrating their various applications in the fields of science and engineering. Chapter 7 explains the processing of composite material from shape memory polymers to cater for newer applications. Chapter 8 is dedicated to the high-temperature applications of shape memory materials, as aerospace applications and modern manufacturing techniques demand thermally stable materials for extreme synthesis and operating environments. Chapter 9 discusses methods of electrical actuation, one of the most modern and demanding actuation techniques. The future prospects of shape memory materials are discussed in Chapter 10, and the book leaves the reader with a fresh overview of the latest technological innovations.

The authors would like to acknowledge Dr. Sreejalakshmi G, associate professor of the Department of Chemistry at the Indian Institute of Space Science and Technology, who was instrumental in providing constructive criticism and has been motivational since the initial stages of the work. The authors are grateful to Dr. Ashok and Shri. Angappan of Air Frame Division, Aeronautical Development Agency, for the valuable input in shaping this work. The reviews and comments of Dr. Usha KM and Dr. Sasikumar P, scientists at the Vikram Sarabhai Space Centre, India, and the comments and input of the reviewers from CRC Press have also been valuable in shaping this book. We express our gratitude to the Vikram Sarabhai Space Centre, Department of Space, India, for helping us to make this publication a reality.

All the references cited are the major sources of data for this book, and hence, all the scientists/researchers behind the cited/referred works are hereby sincerely thanked. A word of acknowledgment is appropriate for our families and all those who offered honest positive criticism and support throughout the publication of this book.

Arun D I
Dr. Chakravarthy P
Dr. Arockiakumar R
Dr. Santhosh B

About the Authors

Arun D I is a scientist at Vikram Sarabhai Space Centre (VSSC) of the Indian Space Research Organization (ISRO), India, and is pursuing his research in the area of smart materials. He has generated several technical reports in the area of smart materials and structures for the aerospace sector.

Dr. Chakravarthy P is a faculty member of the Department of Aerospace Engineering at the Indian Institute of Space Science and Technology. He holds a doctorate in metallurgical and materials engineering, and his current focus is on the development of new materials for aerospace applications. He also had industrial experience in the aerospace sector before taking up his academic assignments.

Dr. Arockiakumar R is a faculty member of the Department of Metallurgy and Material Science Engineering at the National Institute of Technology, Warangal, India. He is a postdoctoral fellow of the National Institute for Materials Science, Japan, and has extensive subject expertise in the field of shape memory materials.

Dr. Santhosh B is a senior scientist at Vikram Sarabhai Space Centre (VSSC) of the Indian Space Research Organization (ISRO). He works with the design and development of composite structures for various launch vehicles, spacecraft, and atmospheric re-entry missions of ISRO space programs. He holds a doctorate in the field of composite structures and had industrial experience before joining ISRO.

1

Introduction

Stone Age, Bronze Age, Iron Age, and Silicon Age are the names given in the timeline of human civilization history based on the materials in use in each period. Each of these periods evolved technologically and culturally to the next era, primarily through advancements in the field of materials science. The significance of each era is portrayed based on the utility of a particular material. The evolution of the ages has witnessed the invention and discovery of newer materials, bringing a change to each era by itself.

The demand for lighter, stronger, and more reliable materials has resulted in the study of a new prospect called *multifunctional materials*. A specific subgroup of such materials with the capability to sense, process, and respond to external stimuli are referred to as *smart materials*. The past few decades have witnessed the development and progression of an extensive array of smart materials for numerous applications in the fields of medicine, mechanics, robotics, aerospace technologies, and so on. Smart materials are those engineered materials that are capable of altering their properties by sensing and responding to environmental conditions. Since the discovery of piezoelectricity (the property of a domain of materials that generate an electrical voltage in response to applied stress) by Pierre and Jacques Curie (1880), the era of smart materials was born and has evolved functionally and theoretically through further studies. Based on the definition of smart materials, a large domain has been identified and categorized as follows: piezoelectric materials, quantum tunneling composites, magnetostrictive materials, light-responsive materials, smart inorganic polymers, halochromic materials, chromogenic materials, photochromic materials, ferrofluids, photomechanical materials, dielectrics, thermoelectric materials, and shape memory materials (SMMs).

The late 1960s witnessed the pioneering concept of synthesizing smart materials for engineering/scientific applications, and these have been used in actuators, vibration damping, microphones, sensors, and transducers. These concepts were primarily based on three approaches, detailed as follows:

- Atomic/molecular-level synthesis of new materials (e.g., doping silicon/gallium to impart semiconducting properties), which emphasizes novel material development targeting specific applications
- Conventional structures with embedded sensors/actuators (e.g., self-healing materials), where smart properties are exploited for existing structural configuration enhancements

- Composite materials with properties superior to their individual components (e.g., shape memory polymer composites [SMPCs], as explained in a later chapter)

1.1 Smart Materials

The phrase *smart materials* refers to those materials, or combinations thereof, that change physically or chemically by sensing and responding to specific environmental stimuli, such as heat, magnetism, electricity, moisture, and so on, and return to their original configuration upon withdrawal of the stimuli.

Nature has always been an inspiration for the development of many stimuli-responsive systems. Engineers have attempted to mimic and develop materials and methods that would artificially respond to such environmental conditions as nature does. The camouflaging of a chameleon or zebrafish, a squid changing its body color to match its surroundings, and the *Mimosa pudica* (touch-me-not) plant responding to the sensation of touch are a few among many known examples.

The outermost layer of the chameleon's skin is transparent, below which are specialized cells called chromatophores that are filled with sacs of different kinds of coloring pigments. Based on body temperature and mood, specific chromatophores expand, releasing specific colors (Figure 1.1a).

The squid has color-changing cells with a central sac holding granules of pigment. The anatomy of the squid is such that this sac is encapsulated by muscles. Contraction of the muscles results in the spreading of the pigment/ink granules around the body, thereby blocking the line of sight of predators (Figure 1.1b).

The unique touch–response property of *M. pudica* has been studied extensively to understand smart traits such as *habituation* and *memory*, which are properties of the Plantae (scientific name for plant) kingdom. Habituation refers to adaptation to the environment and ceasing to respond to nonrelevant biological events, while memory refers to responding to a specific stimulus.

On being disturbed externally by physical touch, various chemicals such as potassium ions are released in certain regions of the plant body. These chemicals are capable of initiating the flow or diffusion of water or electrolyte into or out of cells. This results in a loss of cell pressure. Thereby, the cell collapses, which causes the leaves to close. When this stimulus is transmitted to neighboring leaves, it initiates a process that gives the impression of "touch me not" (Figure 1.2a,b). This is nature's defense mechanism against predators or insects that feed on leaves. Even though this occurs at the expense of energy gained through photosynthesis, the leaves respond smartly to the external stimuli, illustrating a natural smart material concept.

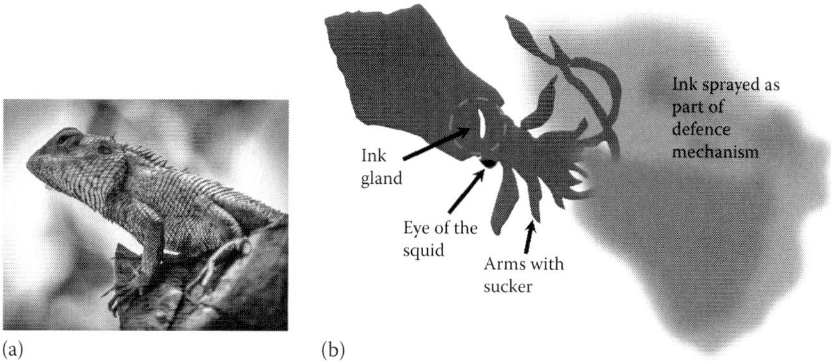

(a) (b)

FIGURE 1.1
Smart responses from nature. (a) Chameleon changes its color corresponding to the environment (courtesy of Whitelily—Pixabay [CC0 Creative Commons]). (b) Squid ejecting ink to escape from its predators.

(a) (b)

FIGURE 1.2
Stimuli response from nature. (a) *M. pudica* plant leaf open. (b) Leaves close in response to touch; the receptors present in the plant's body are activated by an alteration or modification of the plant's shape.

The closed leaves gain back their original open configuration upon the withdrawal of the stimuli. This demonstrates the stimuli-responsive nature of the plant as well as its shape-memorizing ability. Nature provides many such depictions of stimuli-responsive systems that can be adopted to materials/structures/methods used in our daily life. For example, spacecraft antennas are exposed to extreme temperature changes, resulting in dimensional variations (due to expansion or contraction) and its attendant problems. It is inevitable that these materials should have very minimal variations in their dimensions for their best performance. The employment of smart materials for such applications is therefore pertinent. A smart antenna in such a situation shall sense the change in dimensions, judge the correction requirements, and autonomously bring the structure to its best performance. Conventional materials behave in an inert manner, whereas smart materials take over the situation intelligently to provide solutions that overcome the problem. Such smart materials are primarily driven by applications, and the general rules of physics and mechanics are defied, making them unconventional but demanding functional materials. For example, conventional materials when subjected to temperature will become soft and can be molded to the desired shape. Applying the same temperature again shall not bring back the original "predeformed" shape unless acted on by an external force, while many smart materials such as SMMs (as explained in Chapters 2 through 6) are found to behave differently under such situations, thus defying the general rules coined on understanding the behavior of materials so far. Sensing, actuating, and controlling capabilities are intrinsically built into the microstructure of such materials to make a judgment and react to the changes in ambient environment conditions. These stimuli can be changes in temperature, the presence of an electric/magnetic field, moisture, light, adsorbed gas molecules, and pH values, as reported by the scientific community.

The challenges to be addressed are in the selection of the best fabrication methods, the choice of reliable materials for specific functional combinations, and the preparation of materials to a particular density for use in the aerospace and biomedical fields. Material scientists across the world are simplifying the acceptance criteria for smart materials over usual resources, resolving the bottlenecks. They also help in exploiting the properties of smart materials and witnessing a gradual transformation from conventional materials to smart materials in all disciplines.

1.2 Stimuli-Responsive Materials

Let us imagine a window that responds to environmental conditions and adjusts itself in such a way that it provides comfortable living by controlling factors such as ventilation, humidity, and temperature to defined values.

Such a structural object, the "HygroSkin" metereosensitive (abnormally sensitive to weather conditions) biomimetic (the imitation of models, systems, and elements of nature as a solution to complex problems) window pavilion, has been developed by Professor Achim Menges and his team in Orleans, France.

The concept is based on grain direction and the bending stiffness of thin wooden flaps that can absorb air moisture and expand. This opens up the possibility of regulating the perfect living conditions, requiring no human intervention. Figure 1.3a,b depicts the closed configuration and the open configuration for less humid and considerably more humid conditions, respectively, by expanding along the absorbed side to open the thin wooden flaps. Figure 1.3c,d shows the internal and external view of the HygroSkin, respectively, for varied humidity conditions. Thus, by sensing environmental changes, the material responds or actuates to achieve a functional requirement. This is an example of the stimuli-responsive concepts that have evolved as the building blocks of smart structures.

Shape-changing materials are another of the latest trends in the smart material world and have always been an interesting field of study to material scientists. As explained previously, those that can be configured multiple times into a variety of shapes using suitable stimuli can give significant impetus to technological advancement. In line with this, researchers from the University of Pittsburgh's Swanson School of Engineering and from Clemson University in South Carolina have developed a 4D-printed adaptive smart material combining light-responsive fibers with thermoresponsive gels that can serve in soft robotics and other biomimetic applications (Figure 1.4).

The model is of a composite material that integrates a thermoresponsive polymer gel and photosensitive fibers. This material is capable of crawling like a caterpillar upon the application of light and morphs itself when heat is applied. The realization of this model in a laboratory can be of practical use for multi-stimuli-responsive materials that fit various engineering and science applications.

As mentioned previously, multi-stimuli-responsive materials include piezoelectric materials, quantum tunneling composites (QTCs), electrostrictive/magnetostrictive materials, color-changing materials, SMMs, and so on. They occupy the major domain of smart materials.

1.2.1 Piezoelectric Materials

Some materials display spontaneous polarization when subjected to mechanical stress, similar to that of a transducer (devices that can convert one form of energy to another). The appearance of an electrical potential across the sides of a crystal when it is subjected to external mechanical stress is termed *piezoelectricity* or the *piezoelectric effect*. This was first understood from the barium titanate unit crystal, on which the application of stress causes a net dipole moment (Figure 1.5) proportional to the external pressure.

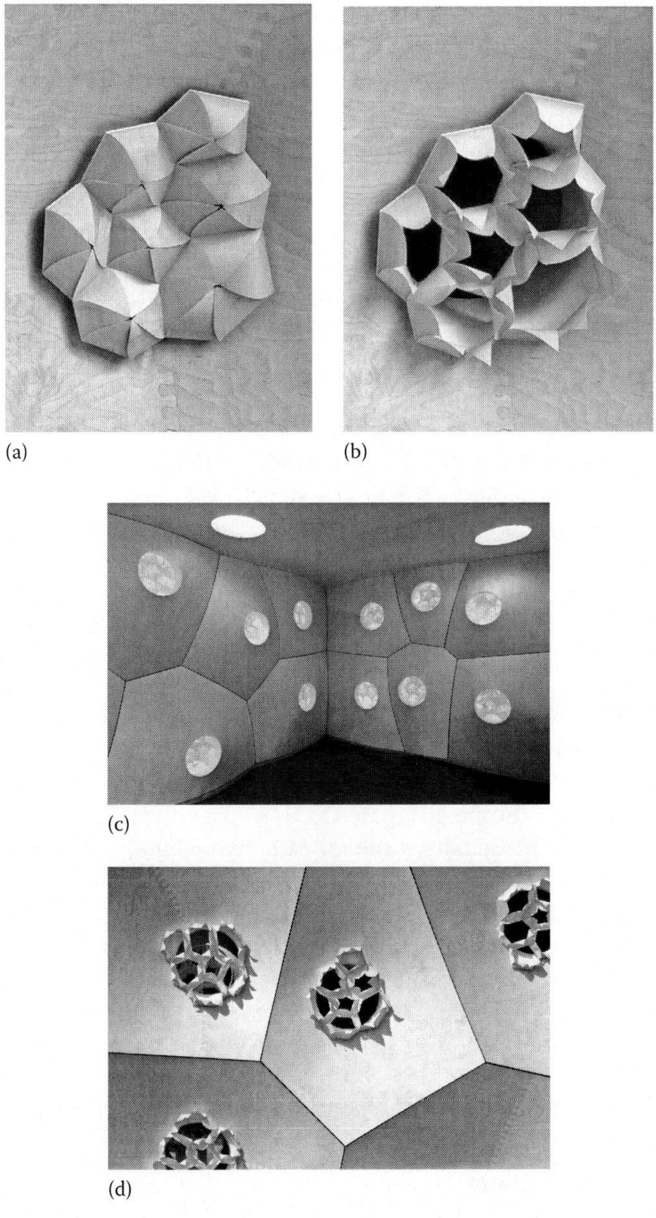

FIGURE 1.3
HygroSkin biomimetic smart structure from smart materials. (a, b) Element-level demonstration of the concept. (c, d) Structure-level demonstration. (From Correa, D., et al., 2013. HygroSkin: A prototype project for the development of a constructional and climate responsive architectural system based on the elastic and hygroscopic properties of wood. In Beesley, P., Khan, O., and Stacey, M. (Eds.), *Proceedings of the 33rd Annual Conference of the Association for Computer Aided Design in Architecture (ACADIA) – Adaptive Architecture*, Waterloo/Buffalo/Nottingham, pp. 33–42.)

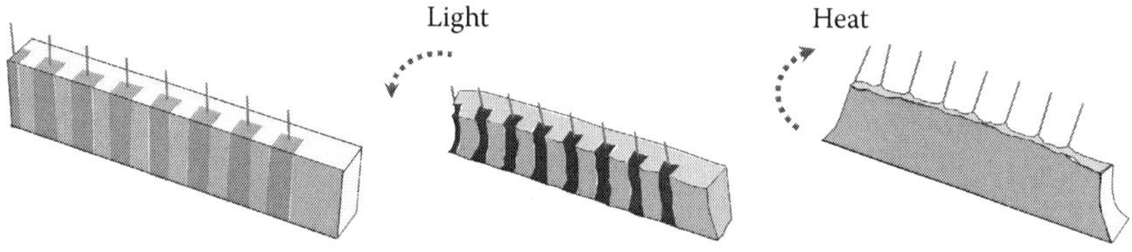

FIGURE 1.4
Shape-morphing material responding to light as well as heat, changing to trained shapes, and recovering the original shape. (Reproduced from Kuksenok, O. et al., 2016 with permission from the Royal Society of Chemistry.)

As depicted in Figure 1.5a, barium ions occupy the corner points of the tetragonal unit cell, while titanium ions are body centered and oxygen ions are face centered. On the application of an external pressure, as in Figure 1.5b, the barium ions are displaced with respect to symmetry and hence a net-effective dipole moment is induced in the unit cell that scales up for the crystal. Figure 1.5c,d shows the change in the net dipole moment due to the application of pressure, thereby inducing an electromotive force across the edges.

The Curie brothers established the first piezoelectric effect by using crystals of tourmaline, quartz, topaz, cane sugar, and Rochelle salt, out of which quartz and Rochelle salt showed the best piezoelectric ability at the time. Sonar devices used during World War I marked the first practical application of piezoelectricity. Research groups in the United States, Russia, and Japan discovered a new class of materials called *ferroelectrics* that are many times better than natural piezoelectric materials due to their faster comparative response. Ferroelectrics exhibit spontaneous electric polarization that is reversed by an external electric field and is analogous to ferromagnetism.

1.2.2 Quantum Tunneling Composites

Another of the most advanced stimuli-responsive materials of interest is QTC. Quantum tunneling is a quantum mechanical process by which minute particles (quantum particles and electrons in a material) pass through a barrier that they generally would not be able to as per the classical theories of physics. This is significant because, if the potential energy barrier is higher than the energy of the particle, the barrier becomes impassible according to classical physics, whereas the theories according to quantum mechanics explain the possibility of the particle making it through. A polymer–metal composite material containing a nonconducting elastomeric binder can be used as a pressure sensor, and this is based on the principle of quantum tunneling. Similar to the piezoelectric mechanism, the application of pressure causes the reduction of resistance to the path of electrons and thus conducts electricity, which otherwise behaves as an insulator in absence of pressure.

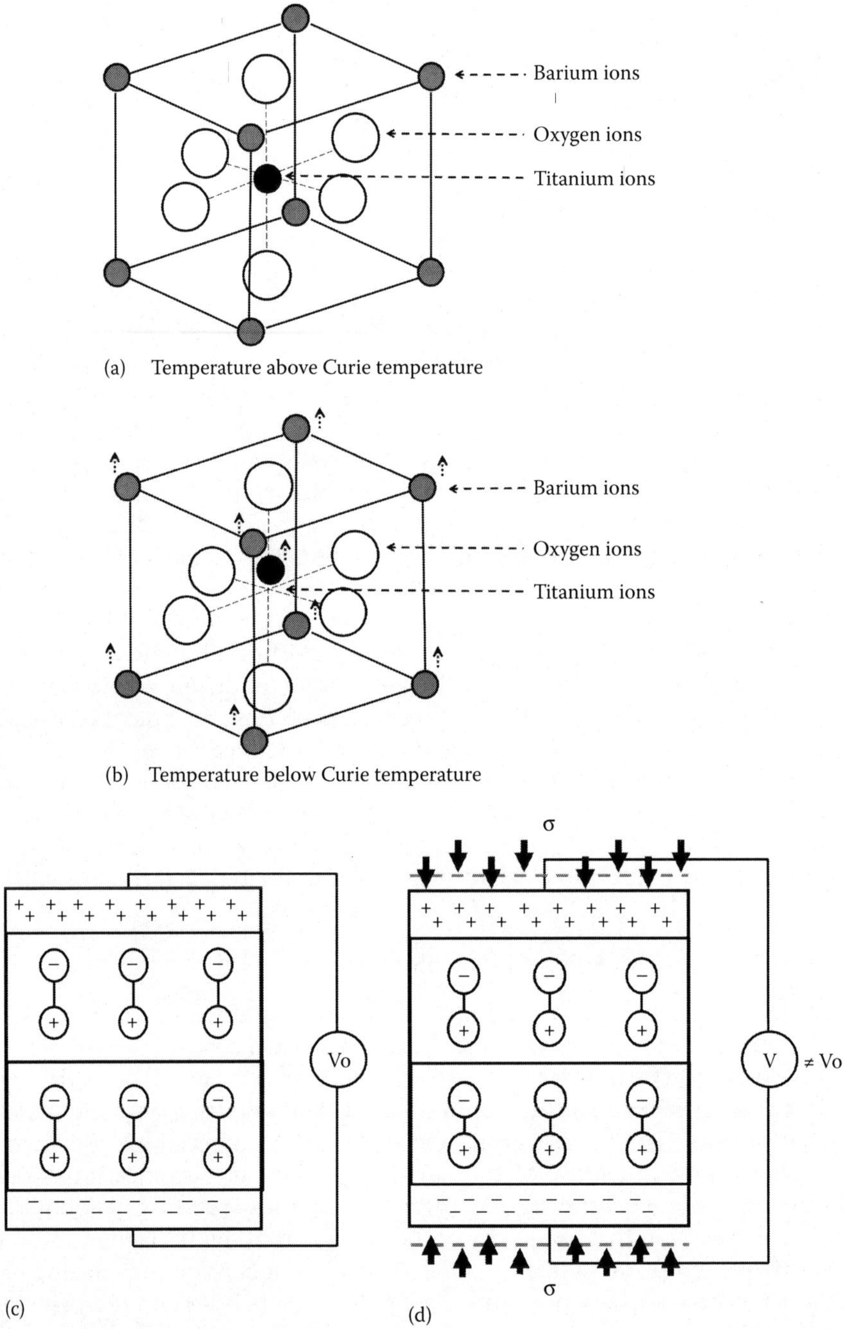

FIGURE 1.5
Depiction of piezoelectric effect and routine application.

Figure 1.6a shows a robot with a QTC attached to the tip of a finger that senses the touch of a human hand. Touch-sensitive surfaces could be a powerful alternative to capacitive touch displays or resistive touch displays, as in the interactive display systems available today (Figure 1.6b). Practical use of this smart material is being explored by the Robonaut project in progress at the National Aeronautics and Space Administration (NASA), where the difficulties of the Robonaut being able to grip something while on a mission are being judged. QTC has found applications in monitoring the blood pressure of patients via electrical signals, feedback for shape morphing in spacecraft/aircraft systems, structural health monitoring in aerospace systems, sports utilities such as fencing masks, and so on.

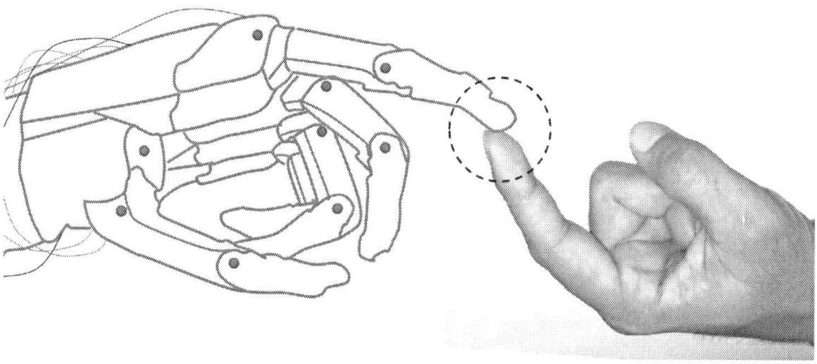

(a)

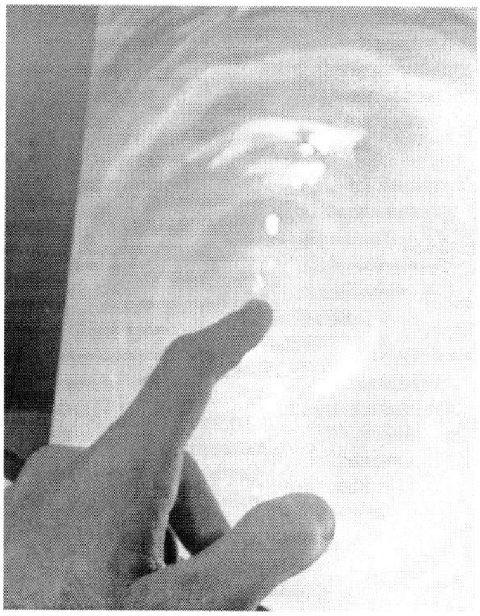

(b)

FIGURE 1.6
Depiction of quantum tunneling composites in real-life applications. (a) Fingertip of a robot equipped with QTC to judge touch. (b) Pressure sensitivity realized for interaction surfaces.

1.2.3 Electrostrictive/Magnetostrictive Materials

Upon exposure to an external electric field, dielectric materials display a shape change with respect to the direction of the field. The reason for this is the accumulation of slight displacements of positive and negative ions in the crystal lattice when an electric field is applied. The accumulated displacement leads to the introduction of a strain corresponding to the field vector. It has been reported that insulating materials with more than one type of atom (which are ionic by nature) show electrostrictive behavior.

Analogous to this, the property of ferromagnetic materials that causes them to change their shape or dimensions due to the influence of an external magnetic field is termed *magnetostriction*. Tiny magnets inside the material realign with the axis of the applied field, resulting in the stretching or shrinking of the bulk. Another interesting effect observed in these materials is the magnetic field that is induced when a mechanical force is applied; this is termed the *Villari effect*.

Both electrostrictive and magnetostrictive materials are highly responsive and are precisely predictable. They are being studied for applications in the field of fuel injection, noise/vibration cancellation experiments, sensing and actuations in aerospace, and in the biomedical field.

1.2.4 Shape Memory Materials

SMMs are special materials that respond by changing shape toward a specific stimulus and memorize the original shape for retrieval. The shape memory effect (SME) was discovered by Arne Olander, a Swedish physicist, in 1932, as the shape-remembering behavior of gold–cadmium alloys. But the discovery of a nickel–titanium shape memory alloy (SMA) by William J. Buehler (1962) of the US Naval Ordinance Laboratory became the turning point in making use of the shape memory property of materials (Ni + Ti + nol = Nitinol). Chapter 2 takes the reader through the mechanisms, characters, and applications of SMAs in detail.

Figure 1.7 shows the shape memory stages of a polytriazole-based polymeric sheet that recovers its original shape (after 40 s) due to the influence of an external stimulus (heat in this case). At the 0 s stage, the shape shown is initially the deformed one, which opens on heating through 10, 20, and 30 s to recover the permanent shape. Thus, the basic concept of this smart material is that a memorized shape is regained when the material is subjected to certain stimuli.

The exploration of SME in alloys has opened new horizons to studies on polymers, gels, and hybrid materials with shape-memorizing abilities, each of which are addressed in different chapters. The authors of this book portray the evolution and establishment of SMM as one of the trending and most widely used smart materials.

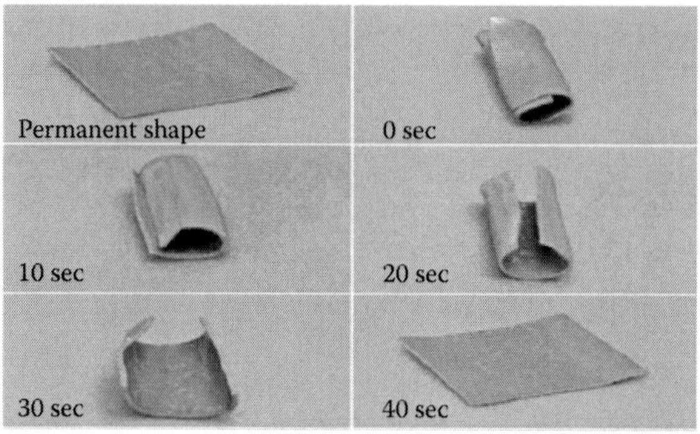

FIGURE 1.7
SME demonstration of a polymeric material (polytriazole-based polymer) under no external stress. (Reproduced from Ragin Ramdas, M. et al., 2015, Synthesis, structure and tunable shape memory properties of polytriazoles: Dual-trigger temperature and repeatable shape recovery, *Journal of Materials Chemistry A*, 3(21): 11596–11606, with permission from the Royal Society of Chemistry.)

The introduction of novel smart materials explains how fascinating mechanisms/systems from nature inspire scientists and engineers to create new materials and structural systems. SMMs are one of the fastest developing fields in terms of stimuli-responsive materials. Such materials are very much required for precise applications such as micro-electro-mechanical devices to replace complex mechanisms and aerospace deployment scenarios.

The development and use of smart materials in various engineering fields has led to increased demand for technological advancements. An increasing trend in demand for smart materials used as motors and actuators (12.89% increase from 2011 to 2016) and structural materials (34.9% increase from 2011 to 2016) has been observed, while the use of transducers has witnessed a decreasing trend (26.87% decrease from 2011 to 2016). The usage of sensors has also decreased with time (21.16% decrease from 2011 to 2016), as depicted in Figure 1.8. The development of such smart materials and their usage in motors and actuators has led to a decreasing trend in the demand of conventional transducers and sensors.

Various smart material developments have been anchored in application-specific demands. Thus, smart materials, specifically stimuli-responsive materials, are broken down further based on their observed behaviors as follows.

Stimuli-responsive materials that change shape in response to a specific stimulus are termed *shape-changing materials*, and if the change is to/from a programmed shape, then the material is termed SMM, the evolution of which is depicted in Figure 1.9 as a flow chart to summarize the chapter. Materials showing SME can be classified as SMAs, shape memory polymers (SMPs), shape memory hybrids (SMHs), shape memory ceramics (SMCs), and shape memory gels (SMGs), which can form composite materials generally termed *shape memory composites* (SMcs).

As introduced earlier, the effect of regaining the original shape of a material by subjecting it to suitable stimuli—be it moisture, heat, light, or chemicals

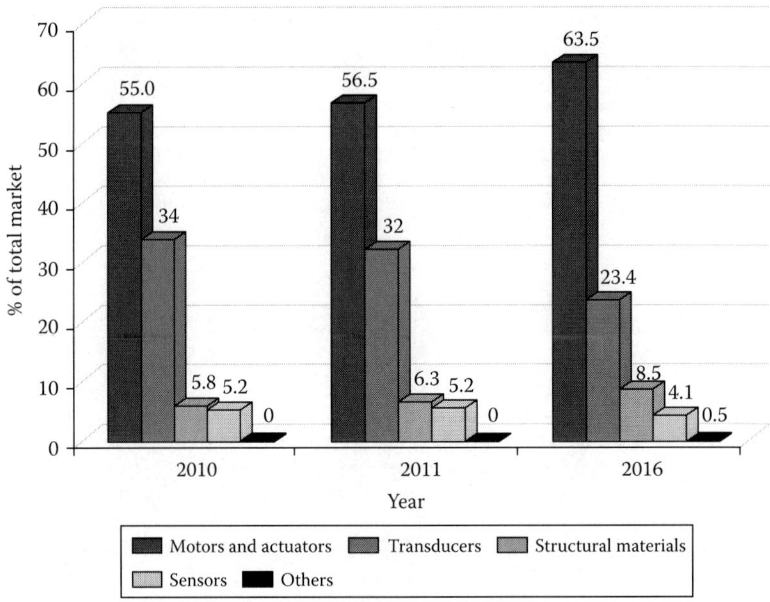

FIGURE 1.8
Global market projection for smart materials during 2010–2016. (Reproduced from Mohd Jani, J. et al., 2014, A review of shape memory alloy research, applications and opportunities, *Materials and Design*, 56, 1078–1113, with permission from Elsevier.)

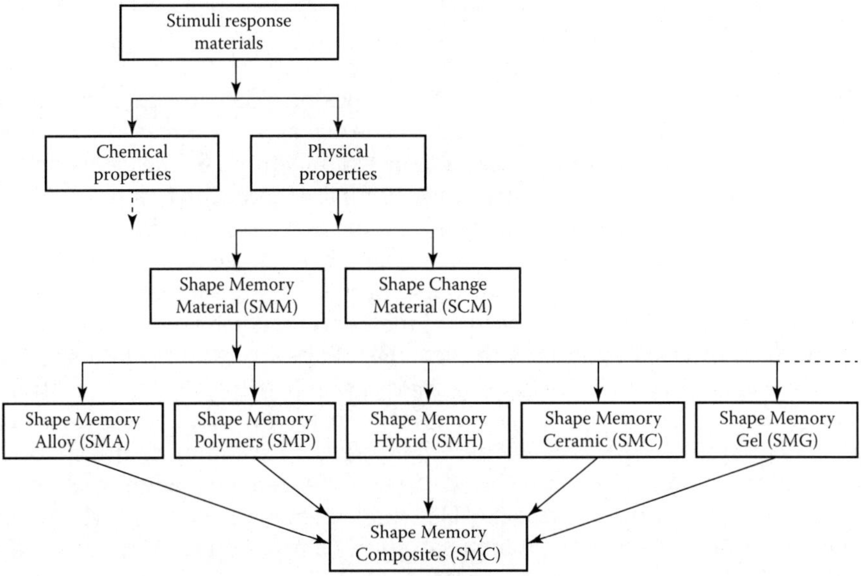

FIGURE 1.9
SMM domain. (Modified from Sun, L. et al., 2012, Stimulus-responsive shape memory materials: A review, *Materials and Design*, 33(1), 577–640, with permission from Elsevier.)

(in the case of polymeric SMMs)—is termed SME. This effect is named after change-inducing stimuli such as hydroactivity, electroactivity, thermoactivity, and magnetoactivity. These distinctive behaviors make them useful in countless engineering and scientific applications, including textiles, deployable structures, shape morphing, cardiovascular surgery, and many other fields.

The previously mentioned SMMs are explained individually in detail in Chapters 2 through 5 (SMAs in Chapter 2, SMCs in Chapter 3, SMGs in Chapter 4, and SMPs in Chapter 5). Chapters 6 and 7 explore SMHs and SMPCs, respectively, where SMMs that form the components of hybrid/composite materials are studied. Chapter 8 deals with the high-temperature applications of SMMs, where alloys and polymers with shape memory behavior at high-temperature operating conditions are explored and compared so that designers can choose application-specific materials based on the study. Chapter 9 concentrates on studies on electrically stimulated SMPs and the methods of achieving electroactivity. Chapter 10 discusses the prospects and futures of smart materials, especially SMPs.

References

Correa, D., Krieg, O., Menges, A., Reichert, S., and Rinderspacher, K., 2013. HygroSkin: A prototype project for the development of a constructional and climate responsive architectural system based on the elastic and hygroscopic properties of wood. In Beesley, P., Khan, O., and Stacey, M. (Eds.), *Proceedings of the 33rd Annual Conference of the Association for Computer Aided Design in Architecture (ACADIA) – Adaptive Architecture*, Waterloo/Buffalo/Nottingham, pp. 33–42.

Kuksenok, O., Balazs, A. C., Liu, Y., Harris, V., Nan, H. Q., Mujica, M., Vasquez, Y., et al., 2016. Stimuli-responsive behavior of composites integrating thermo-responsive gels with photo-responsive fibers, *Materials Horizons*, 3(1): 53–62.

Mohd Jani, J., Leary, M., Subic, A., and Gibson, M. A., 2014. A review of shape memory alloy research, applications and opportunities, *Materials and Design*, 56: 1078–1113.

Ragin Ramdas, M., Santhosh Kumar, K. S., and Reghunadhan Nair, C. P., 2015. Synthesis, structure and tunable shape memory properties of polytriazoles: Dual-trigger temperature and repeatable shape recovery, *Journal of Materials Chemistry A*, 3(21): 11596–11606.

Sun, L., Huang, W. M., Ding, Z., Zhao, Y., Wang, C. C., Purnawali, H., and Tang, C., 2012. Stimulus-responsive shape memory materials: A review, *Materials and Design*, 33(1): 577–640.

Bibliography

Davidson, J. D. and Goulbourne, N. C., 2015. Microscopic mechanisms of the shape memory effect in crosslinked polymers, *Smart Materials and Structures*, 24(5): 055014 (13 pp.).

Fallis, A. G., 2013. Applications of shape memory alloys in space engineering past and future, *Journal of Chemical Information and Modeling*, 59(9): 1689–1699.

Hager, M. D., Bode, S., Weber, C., and Schubert, U. S., 2015. Shape memory polymers: Past, present and future developments, *Progress in Polymer Science*, 49: 3–33.

Han, Z. and Fina, A., 2011. Thermal conductivity of carbon nanotubes and their polymer nanocomposites: A review, *Progress in Polymer Science (Oxford)*, 36(7): 914–944.

Huang, W. M., 2013. Shape memory polymers (SMPs): Current research and future applications, *AZO Materials*: 1–7.

Icardi, U. and Ferrero, L., 2009. Preliminary study of an adaptive wing with shape memory alloy torsion actuators, *Materials and Design*, 30(10): 4200–4210.

Javid, S. and Vatankhah, M., 2016. Fundamental investigation on the mechanisms of shape memory polymer reversibility, *International Journal of Chemical Studies*, 4(4): 43–45.

Leng, J., Lu, H., Liu, Y., Huang, W. M., and Du, S., 2009. Shape-memory polymers: A class of novel smart materials, *MRS Bulletin*, 34(11): 848–855.

Leng, J. and Ye, L., 2009. Smart materials and nanocomposites: Bring composites to the future, *Composites Science and Technology*, 69(13): 2033.

Liu, C., Qin, H., and Mather, P. T., 2007. Review of progress in shape-memory polymers, *Journal of Materials Chemistry*, 17: 1543.

Luo, X. and Mather, P. T., 2013. Design strategies for shape memory polymers, *Current Opinion in Chemical Engineering*, 2(1): 103–111.

Michaud, V., 2004. Can shape memory alloy composites be smart?, *Scripta Materialia*, 50(2): 249–253.

Otsuka, K. and Wayman, C. M., 1999. *Shape Memory Materials*, Cambridge University Press, Cambridge, UK.

Ralls, K. M., Courtney, T, H., and Wulff, J., 1976. *Introduction to Material Science and Engineering*, Wiley, New York.

Sunitha, K., Santhosh Kumar, K. S., Mathew, D., and Reghunadhan Nair, C. P., 2013. Shape memory polymers (SMPs) derived from phenolic cross-linked epoxy resin via click chemistry, *Materials Letters*, 99: 101–104.

Van Humbeeck, J., 1999. Non-medical applications of shape memory alloys, *Materials Science and Engineering: A*, 273–275: 134–148.

Wang, X. L., Oh, I. K., and Kim, J. B., 2009. Enhanced electromechanical performance of carbon nano-fiber reinforced sulfonated poly(styrene-b-[ethylene/butylene]-b-styrene) actuator, *Composites Science and Technology*, 69(13): 2098–2101.

Wei, Z. G. Tang, C. Y., and Lee, W. B., 1997. Design and fabrication of intelligent composites based on shape memory alloys, *Journal of Materials Processing Technology*, 69(1–3): 68–74.

Xiao, X., Kong, D., Qiu, X., Zhang, W., Zhang, F., Liu, L., Liu, Y., Zhang, S., Hu, Y., and Leng, J., 2015. Shape-memory polymers with adjustable high glass transition temperatures, *Macromolecules*, 48(11): 3582–3589.

Xu, J. and Song, J., 2011. Thermal responsive shape memory polymers for biomedical applications. In Fazel-Razai, R. (Ed.), *Biomedical Engineering: Frontiers and Challenges*, InTech, Rijeka, Croatia, pp. 125–142.

2

Shape Memory Alloys

Alloys that return to their initially defined shape when subjected to stress cycling or thermal cycling are known as *shape memory alloys* (SMAs). SMAs possess the ability to change from one crystallographic structure to another in response to a stimulus in the form of temperature or stress. This change in structure means the material has a specific shape at one temperature or stress level and an alternate shape at another. The two crystallographic structures of SMAs are the low-temperature *martensite* phase and the high-temperature *austenite* phase (also known as the *beta* phase or *parent* phase of the alloy).

The main types of SMAs in use are nickel–titanium (Ni-Ti), copper–zinc–aluminum–nickel, and copper–aluminum–nickel, the most popular of which is Nitinol (Ni-Ti, discovered by William J. Buehler and his team at the US Naval Ordnance Laboratory), which has excellent electrical and mechanical properties, is durable, and is resistant to fatigue. Many comprehensive reviews of Ni-Ti and its applications are available across the world, and it has been observed that the anticorrosive properties of Ni-Ti-based SMAs are particularly good. These materials are highly adaptable and are suitable for gripping similar/dissimilar materials that cannot be joined otherwise (Figure 2.5). The demand for Nitinol in cardiovascular stents for angioplasty and other surgical implants/devices explains its popularity over other alloys with shape memory properties.

2.1 Shape Memory Effect in Alloys

Even after permanent strain at room temperature, SMAs can be returned to their original shape by heating them to a transition temperature, hence the term *shape memory*. Figure 2.1 demonstrates the crystallographic processes involved in shape memory phase changes. At high temperatures, SMAs enter the austenitic phase, and at lower temperatures, they enter the martensitic phase. The term *twinned martensite* is given because the diffusionless/shear transformation creates a twinned structure at lower temperatures. The shape change is not visible macroscopically because the twinned martensite (Figure 2.1b) occupies the same space as the austenite (Figure 2.1a). As a result, the martensite formed is referred to as *self-accommodating martensite* (Figure 2.1c). The alloy can be deformed by

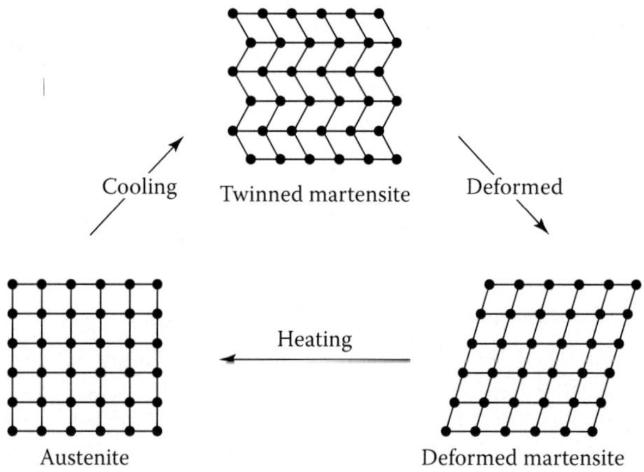

FIGURE 2.1
Schematic of SME: austenite (a) is the high-temperature or parent phase and exhibits a cubic crystalline structure, while martensite (b, c) is the low-temperature phase that exhibits a tetragonal or monoclinic crystalline structure. (Reprinted from Yang, D., 2000, Shape memory alloy and smart hybrid composites: Advanced materials for the 21st century, *Materials and Design* 21(6), 503–505, with permission from Elsevier.)

the application of stress in this phase, prompting the development of the deformed martensite. The same alloy will change back to austenite when heated to a higher temperature. The strain that exists in the deformed martensite will be completely recovered, and the martensite recaptures its unique shape during the process of heating the SMA to the austenite phase. This is called the *shape memory impact* in alloys and describes the behavior of SMAs. The alloys are heated directly or by *Joule heating* (or *ohmic heating*; heating by induction/passing electricity through a conductor) until the transition temperature is reached for the martensite–austenite phase change. The transformation from austenite to martensite may lead to twinned martensite in the absence of internal and external stresses or detwinned martensite if such stresses exist at a sufficient level.

Due to their ability to recover strain in the presence of stress, SMAs are considered *active materials*. As explained earlier, SMAs are found to exist in two different phases, as twinned/detwinned martensite and austenite with reversible transformations within the phases. The martensite phase has a deformation range around 8% and high recovery stress up to 800 MPa if constrained, which are recoverable by heating the SMA to the austenite phase.

The martensitic transformation is a shear-dominant, diffusionless transformation that occurs via the nucleation and growth of the martensitic phase from the austenitic phase. Referring to Figure 2.2, the austenite begins to change to martensite on cooling to the *martensite start* (Ms) temperature of the alloy. The change strictly finishes when it cools to beneath the *martensite finish* (Mf) temperature. On heating, a martensitic–austenitic phase change

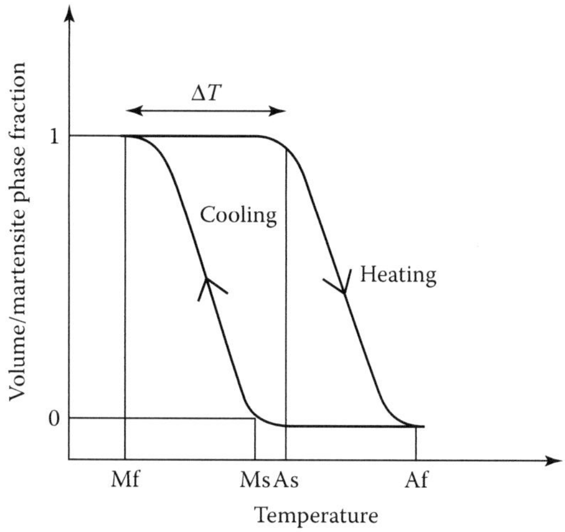

FIGURE 2.2
SMA transformation curve and temperature such that Mf < Ms < As < Af.

is obtained at the *austenite start* (As) temperature. When the alloy is heated to above the *austenite finish* (Af) temperature, the change will end, and beyond the Af temperature, it will be completely austenitic. Austenite and martensite exist together in the middle of the Af and Mf temperatures. There is often hysteresis between the As–Af and Ms–Mf transformation regions, as can be seen on the temperature axis in Figure 2.2.

Research on SMAs has anchored around not only the shape-changing mechanisms but also the embedded SMA elements that could serve as actuators for damping, controlling vibrations, and so on. Damping mainly occurs due to a loss of energy from one form to another during vibratory loads. The energy that is lost may be converted directly into heat or transferred to connected structures or ambient media; this is referred to as *structural damping*. All microscopic and macroscopic changes taking place within the volume of a material or mechanical part and causing energy dissipation during deformation are considered to contribute to material damping.

Intrinsic intelligence (e.g., thermal and stress sensing), actuation, adaptive or active responses, and memory and recovery capabilities are all features of SMAs, which makes them forerunners in aerospace materials.

2.1.1 One-Way and Two-Way Shape Memory in Alloys

Shape memory can be one-way or two-way based on the versatility of the material element and the actuation mechanism. One-way SMAs (OWSMAs) retain their deformed state after the removal of an external force and then return to their original shape upon heating (Figure 2.3a). To begin another shape memory cycle, the martensitic phase must again be distorted. On the

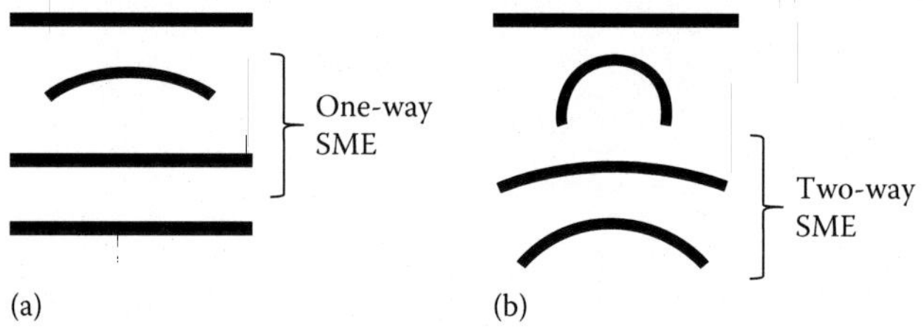

FIGURE 2.3
SMA phases. (a) OWSMA. (b) TWSMA.

other hand, two-way SMAs (TWSMAs), which exhibit the reversible shape memory effect (SME), can remember their shape at both high and low temperatures. In this type of SME, both the austenite and martensite phases are remembered (Figure 2.3b).

Note that the two-way SME (TWSME) response is unfortunately not an inherent property of SMAs and requires repeated thermomechanical treatment (e.g., shape-training cycles) along a specific loading path. Thus, the temporary and permanent shapes held at different temperatures need to be programmed by thermal treatment for the TWSME in alloys.

2.1.2 Superelasticity

Superelasticity or pseudoelasticity in SMAs is a phenomenon whereby the large strains induced by loading an SMA are recovered upon unloading. Superelasticity can be thought of as stress-driven shape memory in SMAs. By utilizing stress above the Ms temperature, the martensite in SMAs can be isothermally induced; this is known as *stress-induced martensite* (SIM). Upon the removal of stress, the shape memory vanishes and the original shape is memorized as an elastic material, which is *mechanical* shape memory rather than *thermal* shape memory. SMAs exhibit superelasticity because of two components: the formation of the reversible stress-induced martensite upon loading the alloy in its austenitic phase and its transformation back to the austenite phase upon unloading.

In Figure 2.4, A denotes the original shape or origin of the effect and *B* indicates the elastic deformation of a specimen at a specific stress level. The elastic deformation of the austenite phase is represented by the line *AB*. Past the point *B*, with further increments in stress, martensite plates start to form. With the further application of stress, the specimen continues to stretch up to point C with no obvious increment in stress level. Here, the sample appears to yield plastically, denoted by the line *BC* (the phenomenon of creep is observed). Be that as it may, in all actuality, the SIM arrangement proceeds up until the austenite–martensite change is complete. Thus, the specimen recovers the strain and returns to its unique measurements along the *CA* line, showing high flexibility or superelasticity.

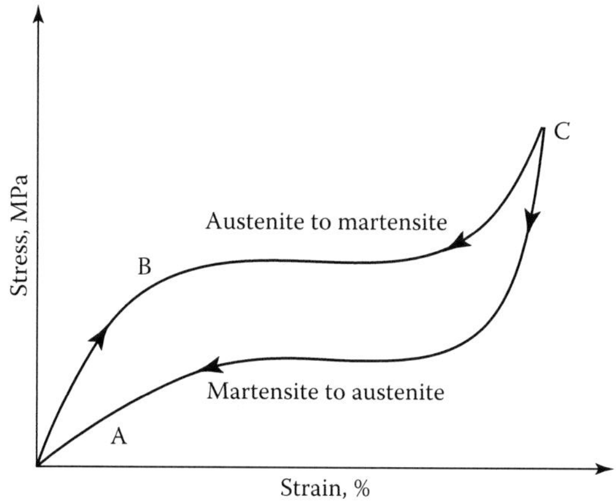

FIGURE 2.4
Superelasticity curve of SMA. Loading curve *ABC* and unloading through *CA*. At *C*, the stress is released and plastic deformation is understood while the unloading curve traces back to the origin.

2.2 SMA Properties and Processing

The demand for smart materials gives many researchers the impetus to take an experimental approach to enhancing the attributes of SMAs by altering their material compositions to give them a wider operating temperature range and better material stability as well as material response.

SMA fibers can be used as fillers in fiber-reinforced composites to improve tensile properties using the residual stress in the matrix. This property is similar to prestressing mechanisms in civil engineering structures, which enhances the tensile/compressive properties of material elements taking heavy loads. Apart from imparting tensile strength as fiber reinforcements, SMA wires can form composites with polymer/metal/ceramic matrices. SMA fibers have useful properties, such as high damping capacities, sensing and actuating functions, and electrical conductivity, thus making them competent candidates for reinforcement in composites. SMA composites have the combined advantages of SMA reinforcements and polymer matrices at a product-level realization.

The *4P relation* developed by Jan Van Humbeeck is a significant contribution to SMA development and evolution. The four Ps are the *principles*, *properties*, *processing*, and *products* required for SMAs to address the challenges at the product level. For property enhancement, the basic principles and processing methods are to be reviewed for optimization, and thus the four Ps are complementary to SMA realization.

Processing complexity, advantageous properties, and availability are the deciding factors in the popularity of any technology, and the same is the case with SMAs. The properties of Nitinol depend on the exact composition of the metal components (the titanium and nickel parts) and the method of preparation. The physical properties of Nitinol include a melting point of around 1240–1310°C and a density of approximately 6.5 g/cc, with an elastic modulus of around 80 GPa for the austenitic phase and 35 GPa for the martensitic phase. The electrical resistivity varies between 75×10^{-6} (martensite) and 80×10^{-6} Ω cm (austenite). A thermal conductivity ranging between 0.086 (martensite) and 0.18 W/cm K (austenite) has been recorded by researchers.

The high reactivity of titanium and the demand for tight compositional control make the processing of Nitinol a challenging task. Two primary melting methods to overcome the challenges are *vacuum arc remelting* (VAR) and *vacuum induction melting* (VIM).

The VAR method uses an electrical arc that is struck between the components of the alloy in a high vacuum in a water-cooled copper mold. The VIM method uses alternating magnetic fields to heat the components in a carbon crucible in a high vacuum. Plasma arc melting, induction skull melting, and e-beam melting are other methods employed in the manufacturing of Nitinol. Controlled heat treatment is known to improve the mechanical behavior of metals as well as alloys. It is important in the case of Nitinol to fine-tune the transformation temperatures and to control the properties of Nitinol products.

Thin Ni-Ti SMA films are fabricated by various physical vapor deposition (PVD) methods such as magnetron sputtering, ion beam sputtering, electron cyclotron resonance (ECR) sputtering, laser ablation, pulsed laser deposition, flash evaporation, electron beam deposition, and vacuum plasma spraying.

2.3 Magnetic SMAs

Similar to thermally responsive SMAs, as discussed previously, SMAs that produce stress and deformations in response to externally applied magnetic fields are known as *magnetic SMAs* (MSMAs). These are magnetostrictive materials that are typically alloys of nickel, manganese, and gallium. In ferromagnetic materials, *magnetically induced reorientation* results in microscopic magnets becoming aligned in the presence of a field. MSMA components are being explored for their possible uses where fast and precise motion is demanded, such as robotics, manufacturing, biomedical elements, sensitive valves, dampers, and so on.

2.4 Applications of SMAs

The ability of SMAs to change shape with a change in temperature has different uses, such as aerospace actuations, medical applications, and some common consumer products such as mobile phone antennas, spectacle frames, and underwire bras.

In the 1970s, SMA coupling designed by the Raychem Corporation was successfully used in Grumman F-14 fighter jets, where the constrained recovery of SMAs was used to join hydraulic lines in the aircraft. This led to pioneering attempts to use SMAs in aerospace applications, where high dynamic loads and constraints in geometric space restrict the use of complex mechanisms.

At low temperatures, rings/couplers of nickel–titanium–iron (Ni-Ti-Fe) alloy are expanded and fitted over the coupling areas of the pipes, as in Figure 2.5a. The ring tries to contract to its original diameter (it "remembers" its original dimensions) when heated to relatively higher ambient temperatures, and thus strong forces are exerted on the pipes, which results in reliable and stable couplings, as in Figure 2.5b. The principle of constrained recovery used in SMAs for coupling pipes has also been used in numerous fasteners and electrical connectors.

Lock-and-release mechanisms (a form of confined recovery) are a common starting point for research into the practical applications of SMAs, which can substitute for the pyrotechnical materials that are commonly used in aerospace deployment or separation systems. This eliminates shock loads during release and thereby reduces the risk of storing an explosive. It is understood that almost 14% of the failed missions before 1984 can be attributed to the shock from the pyrosystems in the spacecraft. The implicit problem associated with pyrotechnics is that they can never be tested before placing them onboard the launch vehicle/spacecraft, whereas shape memory actuation can be tested for any number of cycles before real use. Separation and release mechanisms based on typical shape memory bolts can completely replace pyrosystems. Recent mission failures reported by various space agencies related to separation of heat shields due to malfunction of pyrotechnique

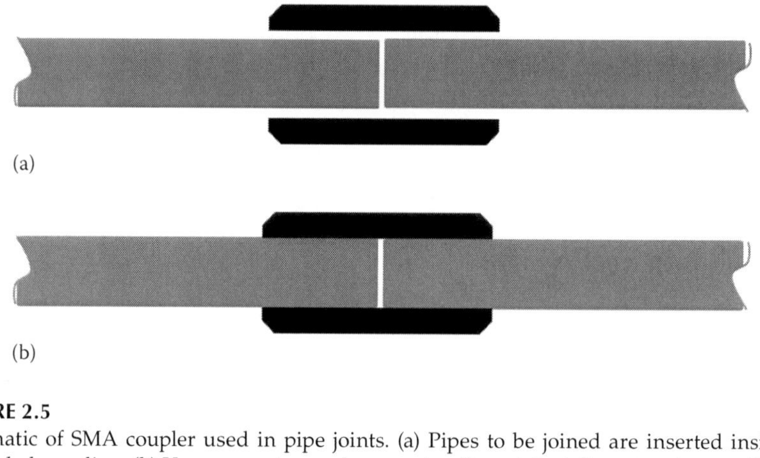

FIGURE 2.5
Schematic of SMA coupler used in pipe joints. (a) Pipes to be joined are inserted inside an expanded coupling. (b) Upon warming and recovering the original diameter, the coupling fits tightly onto the pipes and ensures a reliable metal-to-metal seal. This is used in most advanced aircraft couplings.

and related systems provide new lessons that can lead researchers to accommodate alternate systems such as shape memory devices as replacements for conventional/existing techniques.

The first instance of shape memory material being launched into space was in 1982, when the hermitic durability of SMAs used as thermomechanical couplers was tested in orbit on the Soviet space station *Salyut 7*.

It is a lesser-known fact that the parabolic antennas used by NASA on Apollo missions in the late 1980s were raised using SMA rods.

The Soviet shuttle *Buran* also employed onboard SMA couplers (1988) on an experimental basis. Under Project KRAB, the Rocket and Space Corporation Energia, Russia, studied the deployment of a large space structure using Ti-Ni SMAs, whereas the Sofora and Rapana experiments (the ground tests of experiments designed for space) tested the technique of assembling truss constructions in open space using thermomechanical couplings, the main part of which was an SMA sleeve. Both Sofora and Rapana used electric currents for Joule heating, resulting in activation. In 1994, the Frangibolt family of actuators (a separation system developed by TiNi Aerospace, United States), which are made with Ti-Ni SMAs, was tested for holding the folded solar panels of the NASA spacecraft *Clementine*.

The first time SMAs were tested outside Earth's orbit was when *Sojourner*, the *Pathfinder* rover (1996, NASA), set off to Mars. *Sojourner* had a 30 mm-long Ti-Ni wire, 0.15 mm in diameter, that unshielded a panel with a glass photoelement, which determined the dust content in the planet's atmosphere using differences in radiation intensity from the sun. An SMA actuator used in the Materials Adherence Experiment (MAE) on *Pathfinder* was a landmark achievement, as this took the concept of shape memory beyond Earth's field of influence.

The late 1990s witnessed the use of SMAs in the solar panels of spacecrafts to orient themselves toward the sun. This was based on the efforts of one of the solar sail research groups that began in 1976.

A dexterous robotic hand with movable finger joints was developed for varied space applications, to handle objects of different sizes and shapes and to reduce human interference in assembly, experimentation, and servicing. The actuations and motion were realized by a Nitinol SMA, but it showed less than 9% efficiency, even though the complexity was reduced to a large extent compared with existing systems. Because of their low efficiency, SMAs cannot completely replace conventional actuators.

The NASA Smart Wing program was aimed at optimizing the behavior of lifting bodies. The Smart Aircraft and Marine Propulsion System Demonstration (SAMPSON) of the F-15 flight inlet tested lip skins enhanced with SMA wires for actuation (Figure 2.6b–d). Flight tests in the 16-ft transonic wind tunnel at the NASA Langley Research Center (LRC) (Figure 2.6a) demonstrated the use of SMAs for reducing noise and increasing cruise efficiency by optimizing chevron deflection, thereby controlling the inlet/exit geometry during flight. The chevron outlet of the engine in

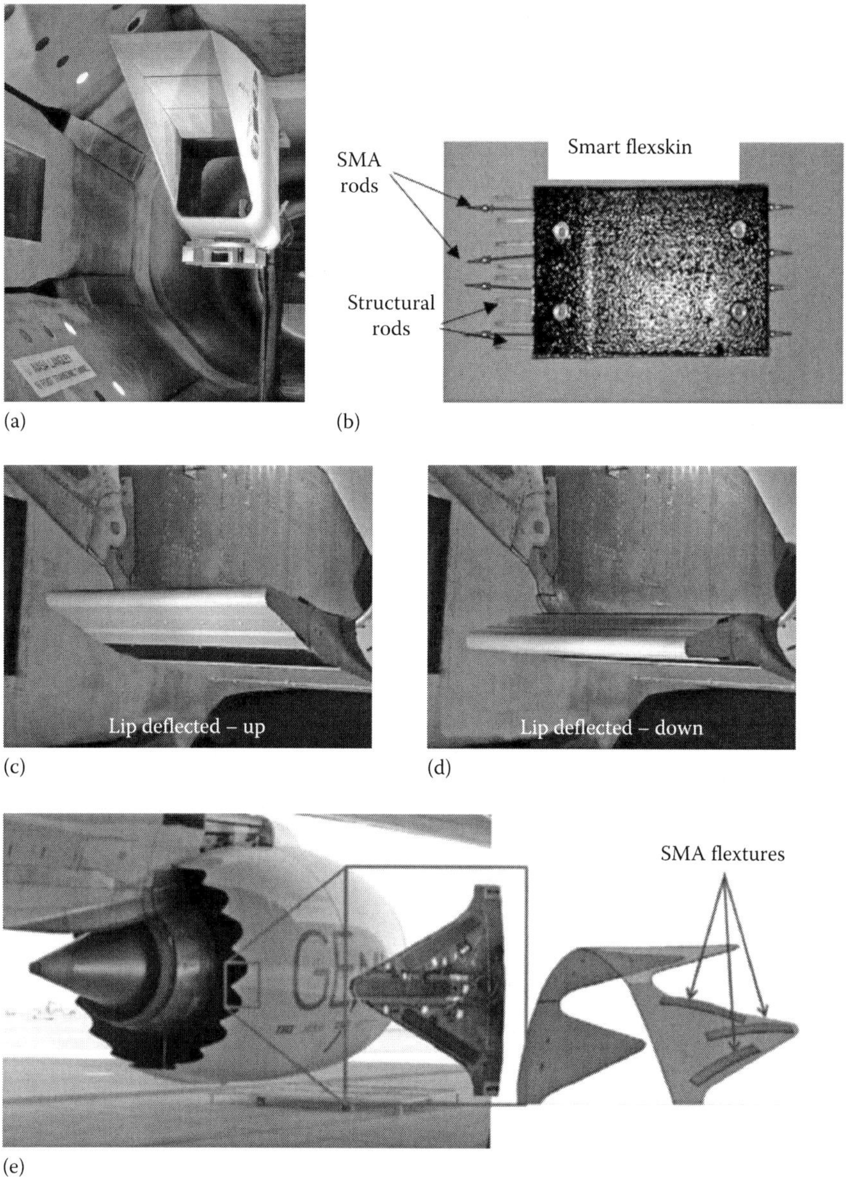

FIGURE 2.6
SMA use in the aerospace field. (a) SAMPSON F-15 inlet cowl in the NASA LRC transonic wind tunnel. (b) Smart flexible skin actuated with SMA. (c, d) Inlet lip deflection during testing. (From Pitt, D. M. et al., 2002, *Proceedings of the SPIE*, 4698: 24–36, with permission.) (e) Boeing variable geometry chevron. (Reprinted from Mohd Jani, J. et al., 2014, A review of SMA research, applications and opportunities, *Materials and Design*, 56: 1078–1113, with permission from Elsevier.)

a Boeing 777-300 ER commercial aircraft is equipped with SMA strips that can change the airflow path, thereby reducing noise during flight (Figure 2.6e).

SMA morphing wings have been tested for aerodynamic loads; forces and torque during flight are assessed by finite-element simulations and modeling. The wings consist of a sandwich box substructure with laminated faces, flexible ribs, and flexible skin, as depicted in Figure 2.7. Thermal stimuli initiate shape changes corresponding to the required lift/drag during flight, which is programmed at the material scale and hence will result in a smart structure.

Chinese material scientists have done an extensive amount of work in shape memory, and the use of SMAs in the sealing connections of the hydraulic system of the Y-12F airplane (1996) was one of the first uses of SMAs in the large-scale domestic aviation field, well before their adaptation into the NASA and Boeing projects.

The multimission Scientific Microsatellite for Advanced Research and Technology (SMART) from the University of Naples, Italy—a piggyback payload in one of the Polar Satellite Launch Vehicle (PSLV) missions—used shape memory wires to deploy antennas. The *Rosetta* spacecraft (2004) of the European Space Agency also used a SMA actuator to release helium for balancing pressure in fuel tanks.

The robotic application of SMAs that are inspired by nature is defined as *biomechanics* (prosthetics, the study of artificial devices that replace missing body parts, is the branch that has benefited most from the research; Figure 2.8a depicts a prosthetic/anthropomorphic hand based on SMA actuation). This field has a direct impact on performance enhancement, from conventional mechanisms to the miniaturization of hardware to provide smaller, faster, more reliable, and more autonomous systems. Clamping

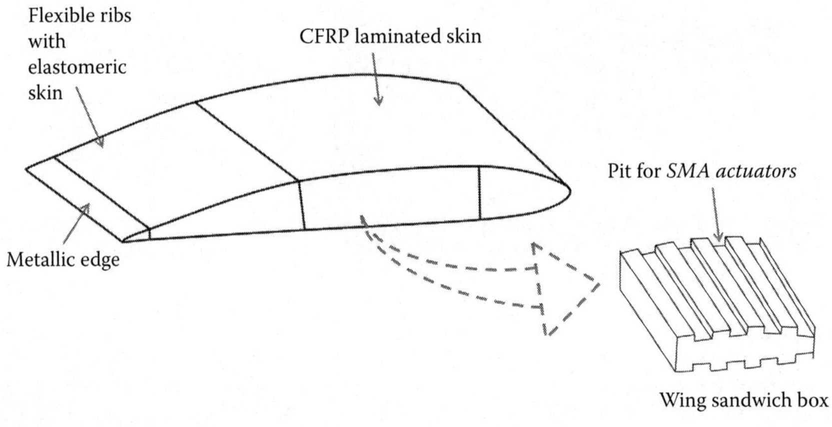

FIGURE 2.7
Adaptive wing structure. (Modified from Icardi, U. and Ferrero, L., 2009, Preliminary study of an adaptive wing with SMA torsion actuators, *Materials and Design*, 30/10: 4200–4210, with permission from Elsevier.)

issues, electrical properties and miniature connections, strain output, and efficiency are issues for which research is ongoing.

BATMAV and BionicOpter (Figure 2.8b) are flying robots developed with SMAs by the Festo Group. The BionicOpter is equipped with four SMA actuators to control with 13 degrees of freedom, which facilitates easy maneuvering and hovering in midair.

Various systems in nature inspire scientists to work on smart materials. Thus, existing complex systems evolve to be more efficient and the complexities die off, resulting in the creation of effortless actuation systems.

Thus, the human body is unarguably one of the biggest spaces in which to develop systems. It provides the liberty to explore, mimic, and learn from its integrity by experimenting on it. Thus, it is even possible to replace/supplement bodily activities with smart materials/actuators. The use of Ni-Ti SMAs in cardiovascular stents for angioplasty is depicted in Figure 2.9.

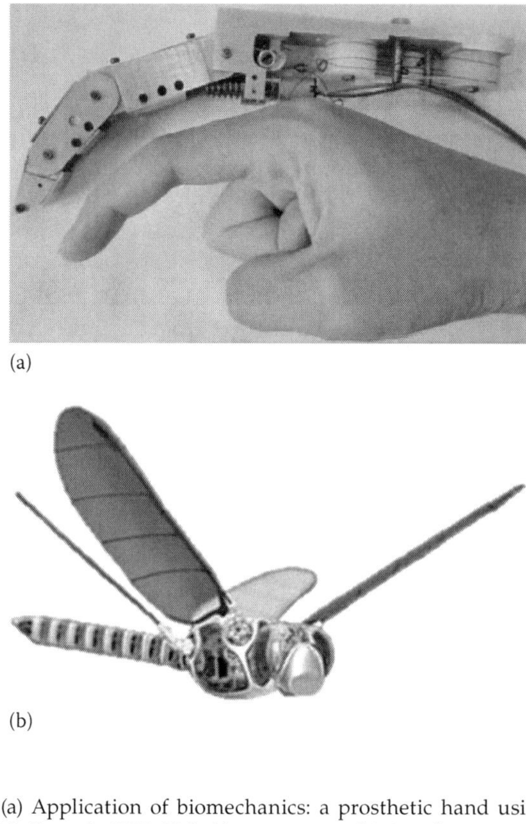

(a)

(b)

FIGURE 2.8
SMA in robotics. (a) Application of biomechanics: a prosthetic hand using SMA actuators. (Reprinted from Fei, G. et al., 2015, Hybrid actuator combining shape memory alloy with DC motor for prosthetic fingers, *Sensors and Actuators A: Physical*, 223:40–48, with permission from Elsevier.) (b) BionicOpter, inspired by the dragonfly. (Reprinted from Mohd Jani, J. et al., 2014, A review of SMA research, applications and opportunities, *Materials and Design*, 56: 1078–1113, with permission from Elsevier.)

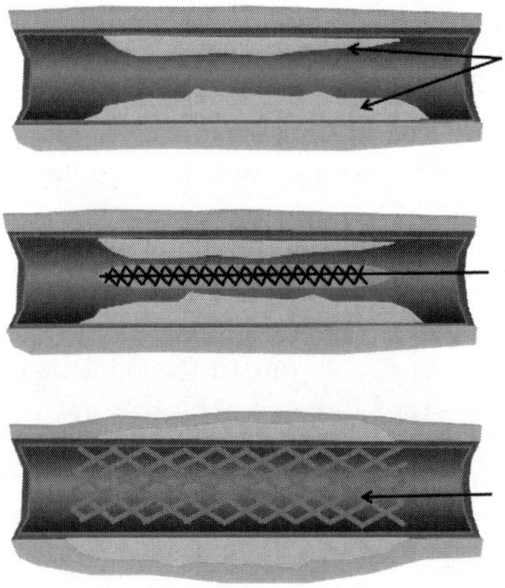

FIGURE 2.9
Deployment of a shape memory cardiovascular stent. A blood vessel filled with plaque, reducing the area for blood flow in the cross-section, has been provided with a stent and a guide wire, which is inserted via some other body part. The stent is deployed to push against the plaque and increase the effective blood flow area, after which the guide wire is removed and the stent remains. (Modified from Mohd Jani, J. et al., 2014, A review of SMA research, applications and opportunities, *Materials and Design*, 56: 1078–1113, with permission from Elsevier.)

The stent is passed through the blood vessel at an accessible point once the block in the blood vessel is identified by angiogram. Upon reaching the clogged portion, the stent is deployed by suitable stimuli, and thus it remains as an inner lining to neutralize deposition and allow the smooth passage of blood.

The maximum possible extent of SMA usage in the human body is depicted in Figure 2.10. Such systems require enormous research before induction as they directly interact with live human body parts. Artificial muscles and implants are the latest advent in the use of SMAs in the biomedical field.

SMAs are being used in a large number of medical applications. For example, broken bones can be mended with SMA staples. The use of SMAs accelerates the healing process, applying compressive forces on the broken bones at the point where the fracture has occurred. Spinal vertebrae spacers prepared with SMAs are another important application in the medical field. These spacers permit the relative motions of two vertebrae to prevent arthrodesis.

Self-healing in shape memory composites (commonly referred to as *polymeric shape memory material*, although SMA systems are also explored) is one of the most discussed abilities as it has applications in the automotive, robotics, biomedical, and aerospace fields.

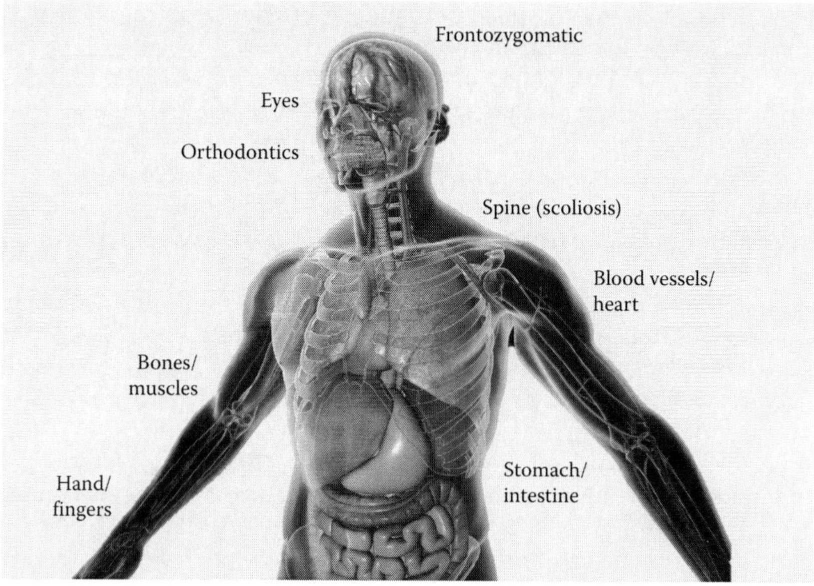

FIGURE 2.10
Existing and potential SMA applications in the biomedical domain. (Modified from Mohd Jani, J. et al., 2014, A review of SMA research, applications and opportunities, *Materials and Design*, 56: 1078–1113, with permission from Elsevier.)

2.5 Challenges in SMAs

SMAs can be considered smart materials in a restricted sense, since they undergo a phase transformation with marked variations in physical or mechanical properties. Another drawback of SMAs for structural applications is their high density (6–8 g/cc for the most common alloys) coupled with their cost. Composite materials containing embedded SMA wires, as discussed earlier in this chapter, are now reaching the point where they can be manufactured in a laboratory at the prototype scale with good reproducibility of the desired properties for simple configurations. The resulting materials are, however, not intrinsically smart, as they often lack adequate sensing potential, apart from temperature. These materials will always be restricted to applications where the frequency of activation is low and where the whole structure, or at least part of it, can be heated during service. In space, the actuator structures are in low-gravity conditions; so, while less power is required, the absence of a convective medium increases the heat transfer problem. Thus, the lack of a cooling system results in low actuation and a slow recovery process. Hence, the applications of SMAs in space demand an efficient cooling system that replicates convection in Earth conditions.

The unaddressed challenges and disadvantages of SMAs pave the way for the introduction of shape memory polymers (SMPs) into the field, to replace or supplement SMAs.

The challenge in designing SMA applications is to overcome their limitations, which include a relatively small usable strain (up to 8%, according to research), low actuation frequency, low controllability, low accuracy, and low energy efficiency.

Self-adaptability, slow response times, phase stability, slow rates of cooling, aging and degradation (durability and reliability on multiple transformation cycles), transformation hysteresis under particular constraints, and low associated energy efficiency (theoretically the maximum energy efficiency of SMAs is in the range of 10%–15%) are phenomena that require further research. Another less addressed issue is the retention of the dimensions of the structures after deployment, as the temperature of outer space swings between −235°C and +150°C. These challenges drive the thought process toward new mechanisms, thus paving the way for studies into shape memory ceramics, gels, hybrids, and polymers.

References

Bosiers, M., Scheinert, D., Hendriks, J., Wissgott, C., Peeters, P., Zeller, T., Brodmann, M., and Staffa, R., 2016. Results from the Tack Optimized Balloon Angioplasty (TOBA) study demonstrate the benefits of minimal metal implants for dissection repair after angioplasty, *Journal of Vascular Surgery*, 64–65: 1552.

Fei, G., Hua, D., and Yi, Z., 2015, Hybrid actuator combining shape memory alloy with DC motor for prosthetic fingers, *Sensors and Actuators A: Physical*, 223: 40–48.

Icardi, U. and Ferrero, L., 2009. Preliminary study of an adaptive wing with shape memory alloy torsion actuators, *Materials and Design*, 30(10): 4200–4210.

Kaplanoglu, E., 2012. Design of shape memory alloy-based and tendon-driven actuated fingers towards a hybrid anthropomorphic prosthetic hand, *International Journal of Advanced Robotic Systems, INTECH*, 9(77): 1–6.

Mohd Jani, J., Leary, M., Subic, A., and Gibson, M. A., 2014. A review of shape memory alloy research, applications and opportunities, *Materials and Design*, 56: 1078–1113.

Pitt, D. M., Dunne, J. P., and White, E. V., 2002. SAMPSON smart inlet design overview and wind tunnel test, II: Wind tunnel test, *Proceedings of the SPIE*, 4698: 24–36.

Yang, D., 2000. Shape memory alloy and smart hybrid composites: Advanced materials for the 21st century, *Materials and Design*, 21(6): 503–505.

Bibliography

Davidson, J. D. and Goulbourne, N. C., 2015. Microscopic mechanisms of the shape memory effect in crosslinked polymers, *Smart Materials and Structures*, 24(5): 055014 (13 pp.)

Hager, M. D., Bode, S., Weber, C. and Schubert, U. S., 2015. Shape memory polymers: Past, present and future developments, *Progress in Polymer Science*, 49: 3–33.

Huang, W. M., Ding, Z., Wang, C. C., Wei, J., Zhao, Y., and Purnawali, H., 2010. Shape memory materials, *Materials Today*, 13(7–8): 54–61.

Huang, W. M., Yang, B., and Qing, F. Y., 2017. *Polyurethane Shape Memory Polymers*, CRC Press, Boca Raton, FL.

Lan, X., Leng, J. S., Liu, Y. J., and Du, S. Y., 2008. Investigate of electrical conductivity of shape-memory polymer filled with carbon black, *Advanced Materials Research*, 47–50: 714–717.

Leng, J. and Ye, L., 2009. Smart materials and nanocomposites: Bring composites to the future, *Composites Science and Technology*, 69(13): 2033.

Michaud, V., 2004. Ca shape memory alloy composites be smart?, *Scripta Materialia*, 50(2): 249–253.

Otsuka, K. and Wayman, C. M., 1999. *Shape Memory Materials*, Cambridge University Press, Cambridge, UK.

Pitt, D. M., Dunne J. P., and White, E. V., 2015. A wind tunnel test of the full scale boeing multi-functional aircraft inlet, multifunctional structures/integration of sensors and antennas, *Meeting Proceedings RTO-MP-AVT-141*, 16: 1–16.

Ralls, K. M., Courtney, T. H., and Wulff, J., 1976. *Introduction to Material Science and Engineering*, Wiley, New York.

Sunitha, K., Santhosh Kumar, K. S., Mathew, D., and Reghunadhan Nair, C. P., 2013. Shape memory polymers (SMPs) derived from phenolic cross-linked epoxy resin via click chemistry, *Materials Letters*, 99: 101–104.

Van Humbeeck, J., 1999. Non-medical applications of shape memory alloys, *Materials Science and Engineering: A*, 273–275: 134–148.

Wang, X. L., Oh, I. K., and Kim, J. B., 2009. Enhanced electromechanical performance of carbon nano-fiber reinforced sulfonated poly(styrene-b-[ethylene/butylene]-b-styrene) actuator, *Composites Science and Technology*, 69(13): 2098–2101.

Xu, J. and Song, J., 2011. Thermal responsive shape memory polymers for biomedical applications. In Fazel-Razai, R. (Ed.), *Biomedical Engineering: Frontiers and Challenges*, InTech, Rijeka, Croatia, pp. 125–142.

Yang, W. G., Lu, H., Huang, W. M., Qi, H. J., Wu, X. L., and Sun, K. Y., 2014. Advanced shape memory technology to reshape product design, manufacturing and recycling, *Polymers*, 6(8): 2287–2308.

3

Shape Memory Ceramics

The shape memory effect (SME) can be observed not only in special alloys or polymers but also in ceramic materials termed *shape memory ceramics* (SMCs). Examples of ceramic materials showing memory properties include partially stabilized zirconia and ferroelectric lead zirconate titanate. In the case of SMCs, rather than stress-induced or thermally induced phase transitions, the elastic strain change associated with the electric field–induced phase transition is utilized to enable smarter actuation than normal piezoelectrics or electrostrictors.

Reyes-Morel et al. (1988) demonstrated SME in ceramics for the first time using CeO_2-stabilized tetragonal zirconia (ZrO_2) polycrystal.

3.1 SME in Ceramics

As explained earlier, SMAs convert heat into mechanical strain by a martensitic phase transformation. SMCs have four different mechanisms of shape recovery: viscoelastic, martensitic, ferroelectric, and ferromagnetic, all of which follow different processes in exhibiting shape memory behavior.

3.1.1 Viscoelastic Mechanism

Mica-based glass-ceramics (40%–60% mica in a glass matrix) exhibit shape recovery of up to 0.5% of strain at high temperatures. The shape is fixed upon cooling and the original shape is recovered upon heating again. This behavior is similar to the hard- and soft-segment melting of polymers, as explained in succeeding chapters. The elastic strain energy due to the deformation strain stored in the glassy phase provides the driving force for shape recovery. The viscoelastic shape memory phenomenon is not limited to mica-based glass-ceramics systems alone, but also includes ceramics and intermetallics such as silicon nitride, silicon carbide, zirconia, and alumina. However, the recovery energy is much smaller than the mica-based glass-ceramics.

3.1.2 Martensitic Transformation Mechanism

Martensitic phase transformation in SMCs is shown by ceramics based on zirconia or partially stabilized zirconia (otherwise called *martensitic ceramics*; mostly ZrO_2 ceramics), where the phase transition is between a tetragonal and a monoclinic structure, which is induced thermally or by the application of stress (thermoelastic transformation). It is interesting to note that martensitic SMCs exhibit pseudoelastic deformation in addition to SME. On loading at room temperature (cooling the austenitic phase), the shape change is effected by a stress-induced transformation from austenite to martensite, and when the load is removed, the new shape remains. Subsequent heating above the transition temperature causes the martensite to revert to its original shape by retracting the loading path.

3.1.3 Ferroelectric and Ferromagnetic Mechanism

Analogous to the austenite and martensite/twinned martensite phases as explained for SMAs, in ceramics, the ferroelectricity-related transition phases are termed *paraelectric* (PE) and *ferroelectric* (FE)/*antiferroelectric* (AFE) (Figure 3.1).

On applying an electric field, the ions inside the material become polarized, and the electric dipole moments in each crystal unit cell are arranged in parallel in the FE phase and against each other in the AFE phase. During the PE phase, the material is nonpolarized, which upon cooling forms the AFE phase with a specific dipole moment. The application of an external field transforms the AFE phase to the FE phase, and the SMC is thereby deformed, as with the martensite in SMAs. This deformed state is unstable and is based purely on the applied field orientation, as it can revert to AFE upon field reversal. Upon heating beyond the critical temperature, the FE phase gains back its original memorized shape. Thus, for FE SMCs, the phase transition is effected by a switching or reorientation of the polarized domains accompanied by lattice distortion, leading to a linear digital displacement and a net volume expansion.

Paramagnetic ferromagnetism, paramagnetic antiferromagnetism, and reversible transformations accompanied by recoverable lattice distortions are shown by some transition metal oxides and are commonly referred to as *ferromagnetic SMCs*. The mechanism of this transformation is similar to the FE phase transition, except that the material in this case is magnetically active rather than electroactive.

3.2 Advantages of SMCs over SMAs

Compared with SMAs, which possess a strain range of up to 8%, as discussed in the previous chapter, it is reported that SMCs in general begin to crack at about 2% after a few transformation cycles. It has been observed that

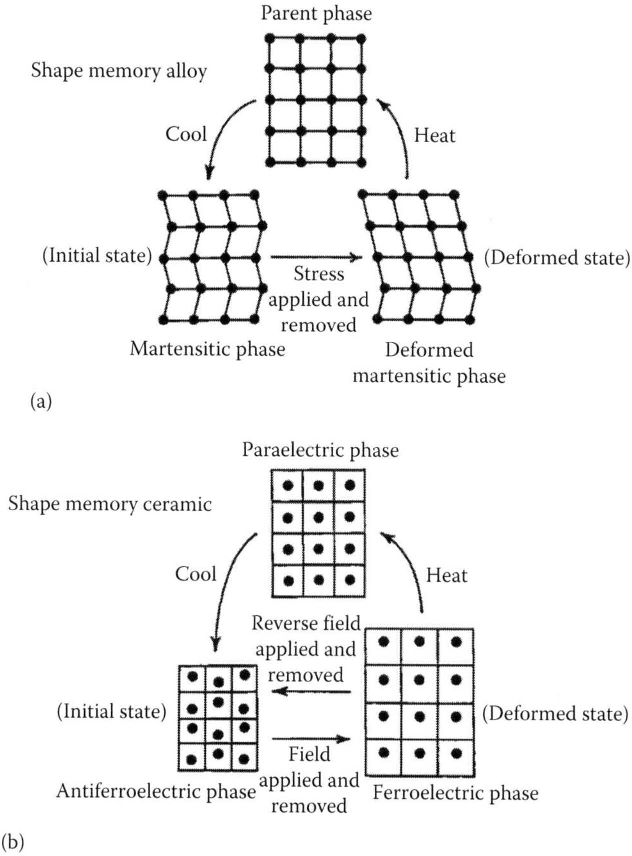

FIGURE 3.1
Comparing SMAs and SMCs. (a) SME in alloys. (b) SMC mechanism. (From Uchino, K., Antiferroelectric shape memory ceramics, *Actuators*, 5: 11, 2016.)

zirconia has a prominent martensitic transformation between the tetragonal and monoclinic phases, with associated shear strains of up to 15%. This is advantageous considering the very high strain range, which permits access to shape memory and superelasticity at higher stress values than SMAs. Ceramics are known to possess high strength, high hardness, and good wear resistance at elevated temperatures compared with metals and alloys, making them the first choice for high-temperature applications. These can also be exploited in shape memory scenarios.

Another important advantage of ceramics over metals/alloys is the range of temperatures in which they can survive and sustain their properties. It is interesting to note that ceramics can be put into a class of potential high-temperature shape memory materials. The most researched SMC, zirconia, is found to exhibit high transformation temperatures ranging between 0°C and 1200°C (Figure 3.2).

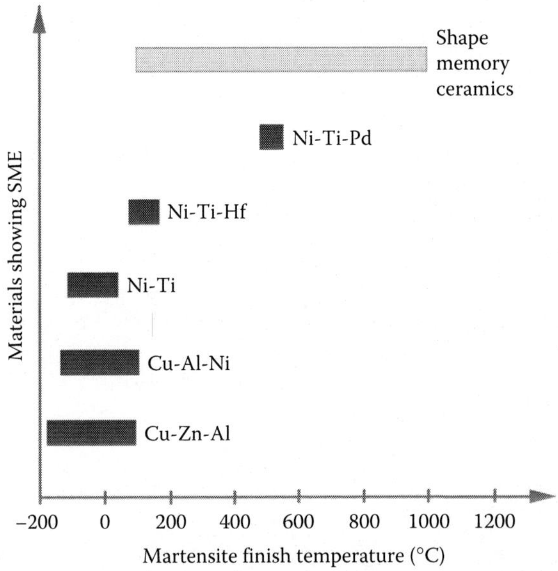

FIGURE 3.2
Operating temperature range comparison of SMAs and SMCs.

Figure 3.2 shows that a combination of copper and aluminum has an operating temperature as low as –200°C, whereas the working temperature of ceramics begins at room temperature. But the maximum temperature in which a nickel alloy with titanium and palladium could operate is ~500°C, whereas the range for SMCs extends up to 1200°C.

3.3 Applications of SMCs

Latching relays and mechanical clampers are potential applications of SMCs, as the ceramic is capable of remaining activated even when electricity is not applied continuously. The schematic in Figure 3.3 shows the structure of a fabricated latching relay, which consists of a mechanical snap-action switch and the shape memory unimorph PZNST. Shape memory unimorph actuators have been manufactured using lead (P) zirconate (Z)–based antiferroelectric ceramics ($Pb_{0.99}Nb_{0.02}[\{Zr_{0.6}Sn_{0.4}\}_{1-y}Ti_y]_{0.98}O_3$). Note that only a short-term electric field pulse (less than 5 milliseconds in duration) can drive the unimorph, which helps to control actuations faster than conventional piezoelectric actuations.

As discussed previously, each type of SMC is unique and exhibits several novel characteristics that are applicable to sensing, actuation, damping,

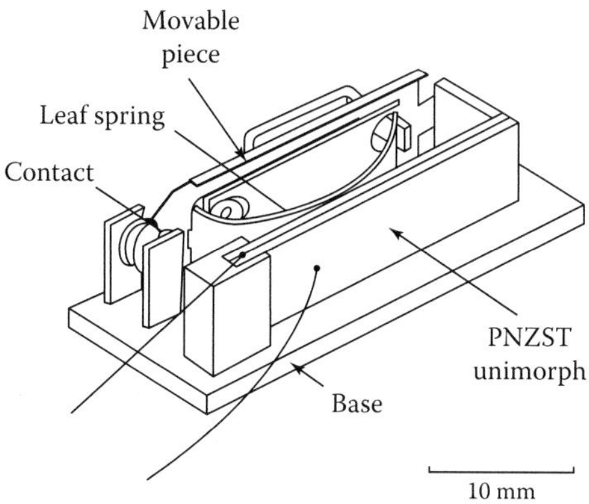

FIGURE 3.3
Latching relay using shape memory ceramic. (From Uchino, K., Antiferroelectric shape memory ceramics, *Actuators*, 5: 11, 2016.)

adaptive responses, shape memory, and superelasticity in smart structures. However, the brittleness of ceramic matrices is a disadvantage and points the research to more flexible materials such as gels and polymers to overcome this shortcoming, even though the application temperatures of ceramics are yet to be confirmed.

References

Reyes-Morel, P. E., Cherng, J. –S., and Chen, I.-W., 1988, Transformation plasticity of CeO2-stabilized tetragonal zirconia polycrystals: II, pseudoelasticity and shape memory effect, *Journal of the American Ceramic Society*, 71 (8): 648–57

Uchino, K., 2016. Antiferroelectric shape memory ceramics, *Actuators*, 5: 11.

Bibliography

Gall, K., 2002, Shape memory polymer nanocomposites, *Acta Materialia*, 50(20): 5115–5126.

Huang, W. M., Yang, B., and Qing, F. Y., 2017, *Polyurethane Shape Memory Polymers*, CRC Press, Boca Raton, FL.

Lai, A., Du, Z., Gan, C. L., and Schuh, C. A., 2013, Shape memory and superelastic ceramics at small scales, *Science*, 341(6153): 1505–1508.

Quan, D. and Hai, X., 2015, Shape memory alloy in various aviation field, *Procedia Engineering*, 99: 1241–1246.

4

Shape Memory Gels

Due to the obvious advantages in material properties, smart materials are taking over conventional materials in many engineering applications. Thus, smart gels are contributing toward a new generation of biomaterials that serve as templates for biomedical devices or scaffolds for tissue engineering, biosensors, and so on, and toward the development of self-healing materials.

4.1 Shape Memory Effect in Gels

A *smart gel* is a material that changes its physical or chemical properties—for example, by swelling or change in light-reflecting capabilities—in response to specific external environments, such as the presence of moisture or changes in chemicals, temperature, or field proximity. Gels are polymeric materials acting as a porous matrix that can accommodate fluids to swell or interact with their surroundings. The types of stimuli that affect smart gels are physical and chemical factors such as changes in temperature, light, electric forces, magnetic forces, and mechanical forces, pH changes or solvent exchanges, and so on. Smart gels generally expand/swell or contract in response to stimuli as a result of changes in the structural arrangement of the polymeric entity. If the smart gel can return to its original shape on revisiting the stimuli, then it can be called a *shape memory gel* (SMG).

The addition of nano-/microparticles that bestow the property of shape memory or shape morphing to gels is being studied, in line with other materials that exhibit the shape memory effect (SME).

The Soft and Wet Matter Engineering Lab (SWEL) at Yamagata University, Japan, has developed an SMG that is soft and wet, with properties such as high water absorbance, softness, transparency, and extremely low friction, which could be advantageous in biomedicine, aerospace, and various other fields.

The Department of Engineering Science at Oxford University, United Kingdom, are working on a liquid suspension of very small particles termed *sol* as an advanced form of stable gel with smart properties. These materials possess the ability to shift shape from gel to sol and the reverse, responding to certain external stimuli. Sol represents a colloidal state where the particles

(a) (b) (c)

FIGURE 4.1
SME in hydrogel temporary shape trained to respond and regain original shape and recovered original shape by moisture absorption. (Reprinted from Huang, W. M. et al., 2013, Shaping tissue with shape memory materials, *Advanced Drug Delivery Reviews*, 65(4), 515–535, with permission from Elsevier.)

appear as discrete elements in a larger system volume, which results in a more flexible internal structure that lowers the transition temperature, thereby giving the gel its shape memory behavior.

Microelectromechanical/nanoelectromechanical systems (MEMS/NEMS), self-healing coatings, electrolytes for rechargeable batteries, or enhanced dielectrics for supercapacitors could be potentially suitable applications for such shape-shifting/-memorizing gels.

Microfluidic devices are being synthesized using SMGs. Hydrogels with the SME property serve to make microvalves and flow modulators, which have advantages such as relatively simple fabrication, no external power requirement, no integrated electronics, and large displacement force generation.

A temperature-stimulated SMG hydrogel is depicted in Figure 4.1 with its temporary and recovered shapes. Figure 4.1a shows the original shape, which is recovered in Figure 4.1c with suitable stimuli. Here, the stimulus is moisture, causing the gel to swell and regain its shape from a temporary shape, shown in Figure 4.1b.

Studies have reported that electroactive or thermally stimulated hydrogels are being synthesized for greater precision in biomedical operations. These hydrogels are porous materials that swell with water absorption, with pore sizes ranging from less than 1 μm (Figure 4.2a) to larger than 1 mm (Figure 4.2b), as evident from the scanning electron micrographs shown in the figure.

4.2 Applications of SMGs

The major trending applications of smart SMGs are artificial muscle research and drug delivery/release systems. For example, in drug delivery applications, the patient is injected with a smart gel containing a water-soluble drug.

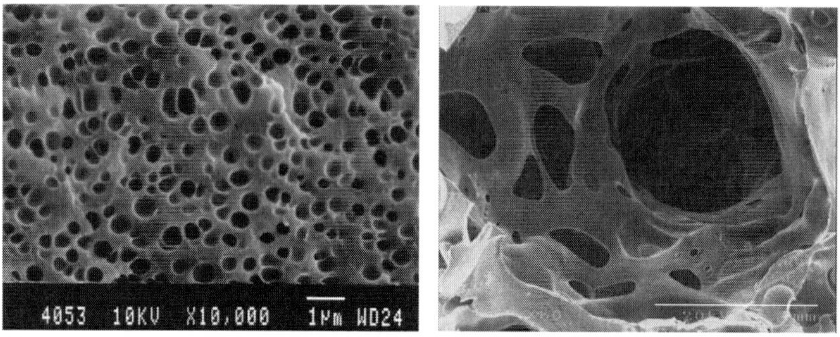

(a) (b)

FIGURE 4.2
SEM image of pores in a hydrogel for pores of different sizes in same magnification scale. (Reprinted from Chaterji, S. et al., 2007, Smart polymeric gels: Redefining the limits of biomedical devices, *Progress in Polymer Science*, 32/8–9, with permission from Elsevier.)

Upon activation by a stimulus, which can be either temperature or pH, the gel expands and allows blood/salt to enter and wash out the drug in specified environments. Using smart gels, it is possible to accurately release drugs at a specified speed to tumors, specific body parts that are otherwise inaccessible, or other areas requiring precision drug delivery. Some gels are capable of mimicking the role of muscles in the body in response to electrical signals sent from the brain.

The National Institute of Standards and Technology (NIST), US Department of Commerce, is studying *shake gels*, which transform into gels when shaken and then relax back into the liquid phase after some time, and have applications in shock absorbers.

Structural applications drive research toward advanced or enhanced materials that may be a combination of multiple intelligent materials. The following chapter will discuss polymeric materials with shape memory properties, which prove to be superior to gels in many aspects.

References

Chaterji, S., Kwon, I. K., and Park, K. 2007, Smart polymeric gels: Redefining the limits of biomedical devices, *Progress in Polymer Science*, 32(8–9): 1083–1122.

Huang, W. M., Song, C. L., Fu, Y. Q., Wang, C. C., Zhao Y., Purnawali H., Lu H. B., Tang C., Ding Z., and Zhang J. L., 2013. Shaping tissue with shape memory materials, *Advanced Drug Delivery Reviews*, 65: 515–535.

Bibliography

Chaterji, S., Kwon, K., and Park, K., 2005. Smart polymeric gels: Redefining the limits of biomedical devices, *Biophysical Chemistry*, 257(5): 2432–2437.

Huang, W. M., Yang, B., and Qing, F. Y., 2017. *Polyurethane Shape Memory Polymers*, CRC Press, Boca Raton, FL.

Huang, W. M., Ding, Z., Wang, C. C., Wei, J., Zhao, Y., and Purnawali, H., 2010. Shape memory materials, *Materials Today*, 13(7–8): 54–61.

Van Humbeeck, J., 1999. Non-medical applications of shape memory alloys, *Materials Science and Engineering: A*, 273–275: 134–148.

5

Shape Memory Polymers

Polymers that are able to "memorize" a permanent shape and can be manipulated so that a certain temporary shape will be "fixed" under appropriate conditions are called *shape memory polymers* (SMPs). The transformation is stimulated by temperature changes across the transition temperature of the polymer.

5.1 Shape Memory Effect in Polymers

We have studied the properties of shape memory alloys (SMAs), shape memory ceramics (SMCs), and shape memory gels (SMGs) in previous chapters, and comparing their respective advantages, SMPs are superior in many ways, which will be explored in detail in this chapter. Their enhanced usable strain adds to the versatility of forming composites, which makes them stand out among other shape memory materials. Polymer structures and morphology together are responsible for the shape memory effect (SME), along with proper shape-training techniques. SMPs are extensible up to 800% (as suggested by the experimental data in the published literature [Liu et al., 2007]), and their exclusive features are tunable elastic moduli, tunable response temperatures, and low mass densities. These findings are reinforced by works on carbon-filled SMP composites and fiber-reinforced SMPs, which will be discussed in Chapter 7. Diverting the lion's share of the research from alloys to polymers will take us through their myriad advantages, such as low weight, low cost, large strain (comparatively), and shape changes under external stimuli, such as heat, light, electricity, magnetic fields, solvents, pH, and so on. Before going into the details, let us understand the basics of shape memory in polymers and how it differs from that in other materials. Figure 5.1 depicts the complete cycle of an SMP from heating to a transition temperature (similar to SMAs), deformation, shape fixing, unloading, and shape recovery.

The curve starts from point (iv), where the temperature of the material is brought to Tg, which is the glass transition temperature (the temperature where glassy substances become soft and rubbery). The rising portion along the stress axis denotes the shape-training process from point (iv) to point (i). The material is given its temporary shape at point (i) and is then cooled along the line (i)–(ii). On reaching point (ii), the temporary shape is

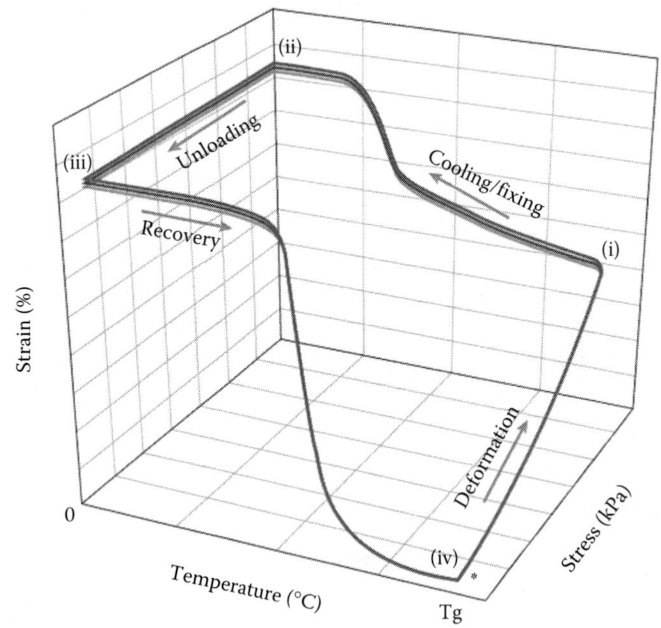

FIGURE 5.1
Three-dimensional thermomechanical cycle of a typical polymeric shape memory material. (Adapted from Luo, X. and Mather, P. T., 2010, Conductive shape memory nanocomposites for high speed electrical actuation, *Soft Matter*, 6(10): 2146, with permission from the American Chemical Society.)

fixed and the stress/external loading can be removed, returning along the stress axis through the line (ii)–(iii). This is the operating temperature of the polymer. The polymer will again be heated to its transition temperature at point (iv) (here, Tg) when required to release the strain and return to its original shape.

Figure 5.2 shows the original/recovered and temporary shapes of a polyurethane SMP.

Going back to the early 1940s, Vernon et al. (1941) claimed to discover "shape memory" in a US patent on dental biomaterials (methacrylic ester resin), which has evolved over decades to one of the most significant smart materials of the present day.

5.2 Mechanism of Shape Memory in Polymeric Materials

Application-driven industrial research is now complemented by academic research, which has begun to investigate the underlying mechanisms and design principles of these materials. The mechanism of SMPs is totally

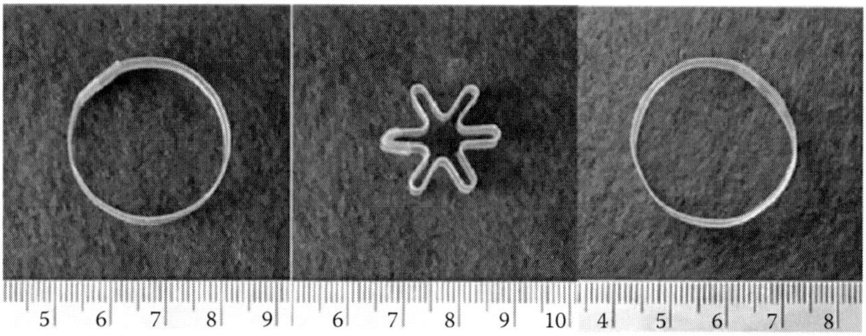

FIGURE 5.2
SME in a polyurethane stent. (Reprinted from Sun, L. et al., 2012, Stimuli-responsive shape memory materials: A review, *Materials and Design*, 33: 577–640, with permission from Elsevier.)

different from the martensitic phase transition of SMAs or ferroelectric phase transition of SMCs. Here, shape shifting and recovery are achieved through a polymeric segmental phase transition. The existence of two phases inside polymers explains this process, one allowing for shape fixity (hard segment/net points/frozen portion) and the other allowing for reversibility or recovery (soft segments). These phases are primarily held responsible for maintaining dimensional stability and determining the reversible thermal response of the polymer, respectively. Polymer shape memory occurs due to glass transition (the temperature at which the phase changes from a rigid to a flexible, soft phase) or the melting transition of a hard segment to a soft segment.

SME can be interpreted as an efficient method of energy storage: the initial heating and training to store energy and the recovery of the shape to dissipate it (Figure 5.3). Thus, achieving a temporary shape or shape fixity can be referred to as the *unstable* form of the material, which stores energy as deformation (as in the predeformed shape in Figure 5.3). Upon heating the material above its shape transition temperature, it will retrieve the original shape by giving away the stored potential.

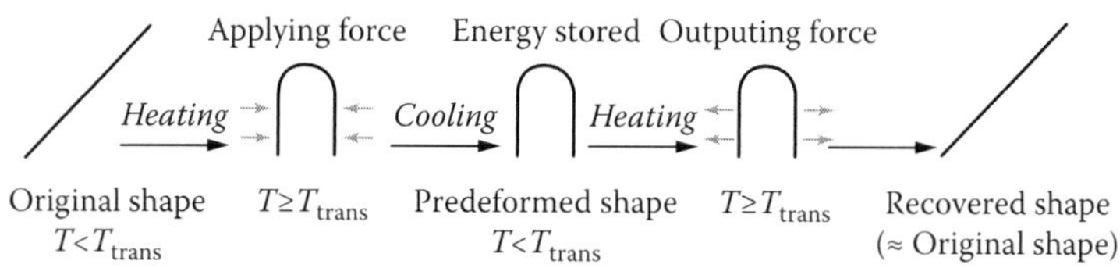

FIGURE 5.3
Schematic of SME during a typical thermomechanical cycle. This can be applicable to any type of material, from alloys to polymers. (Reprinted from Leng, J. et al., 2011, Shape memory polymers and their composites: Stimulus methods and applications, *Progress in Material Science*, 56(7): 1077–1135, with permission from Elsevier.)

The ability of the reversible segments in a polymeric material to retain their applied mechanical deformation is termed *shape fixity* (R_f), whereas *shape recovery* (R_r) denotes the ability of the material to recover its original shape. R_f and R_r are specifically calculated to a particular SMP with Equations 5.1 and 5.2, respectively.

The shape-fixing ratio R_f (representing the stability of the SMP in its strained temporary shape) is defined as the ratio of the deformation after unloading and the deformation at the fixing temperature under the load σ, and is expressed as

$$R_f(N) = \left(\varepsilon_u(N)/\varepsilon_m(N)\right) \times 100\% \tag{5.1}$$

where:

ε_u is the strain in the fixed temporary shape
ε_m is the strain after the stretching step (before cooling)
N is the number of cycles

Referring to Figure 5.1, the portion between (ii) and (iii) should ideally be a straight line parallel to the stress axis for perfect shape fixity (i.e., $\varepsilon_u = \varepsilon_m$).

The shape recovery ratio R_r (representing the efficiency of the recovery of an SMP to its original shape) is defined as the ratio of the recovered deformation at the recovery temperature and the fixed deformation under stress at the deformation temperature, and is expressed as

$$R_r(N) = \left(\varepsilon_m(N) - \varepsilon_p(N)\right) / \left(\varepsilon_m(N) - \varepsilon_p(N-1)\right) \times 100\% \tag{5.2}$$

where ε_p is the strain after recovery.

Referring to Figure 5.1, the recovery line starting from (iii) should meet (iv) perfectly, indicating efficient shape recovery.

Research on SMPs is widespread across their various properties, the structure of the polymeric system, and the factors influencing SME, such as transition temperature, cross-linking, and so on. Polynorborene (developed by the French company CdF Chimie and commercialized by the Japanese company Nippon Zeon in 1984) was the first SMP developed with a transition temperature ranging between 35°C and 40°C. The significance of a transition temperature of this range is that it can be of bioclinical use if proven compatible with the human body. During surgery, suture threads with shape memory properties are stretched beyond their normal length at their transition temperature and then cooled below normal body temperature. Since soft flesh cannot take the tension at the incision, the knots made will be loose during the stitching process. The sutures will be left to reach body temperature (the stimuli) and recover their original unstretched length, tightening the stitch. Thus, biocompatible shape memory materials are especially useful in biomedical engineering.

Entropy elasticity, a property of SMPs that releases polymeric chains from their strained configuration (the temporary trained shape) to a less complex configuration (the original shape as memorized), acts as an impetus for shape memory realization.

The basic molecular composition of SMPs and their active cross-linked architectures ensures that they maintain a stable shape at the macroscopic level until they are acted on by triggering stimuli and change to their temporary/original shapes.

This configuration is a three-dimensional molecular network formed by cross-linked polymer net points, providing stability in both the original and recovered shapes. The cross-linking domain can be either physical, which is achieved by forming crystalline or glassy phases (Figure 5.4a), or chemical, where the individual polymer chains are linked by covalent bonds (Figure 5.4b).

The three basic working mechanisms of SMPs are as follows:

1. *Dual-state mechanism (DSM)*: The molecular chains are the reversible phase and the nodes of macromolecule segments (analogous to matrices in composites) attributed to physical or chemical cross-links (Figure 5.4) are the permanent phase.
2. *Dual-component mechanism (DCM)*: One component is elastic and the other is able to switch stiffness, depending on the stimuli.
3. *Partial-transition mechanism (PTM):* Instead of a complete transition, heating may stop at a temperature within the transition range. As such, when deformed, the unsoftened portion may serve as the elastic component to store elastic energy, while the softened portion may behave as the transition component.

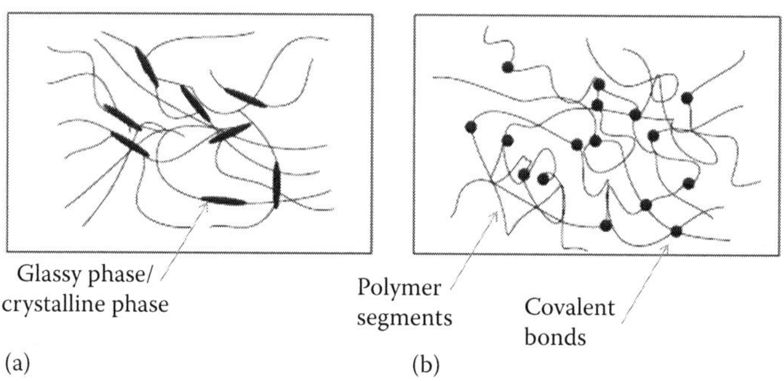

FIGURE 5.4
Three-dimensional polymer networks. (a) Physical cross-linking. (b) Chemical cross-linking. (Reprinted from Leng, J. et al., 2011, Shape memory polymers and their composites: Stimulus methods and applications, *Progress in Material Science*, 56(7): 1077–1135, with permission from Elsevier.)

5.3 Transition Temperature

Phase changes are achieved by training the polymeric material at a specific high temperature defined as the *transition temperature* (T_{trans}) or *glass transition temperature* (T_g). The T_g of a material characterizes the range of temperatures over which this glass transition occurs. It is always lower than the melting temperature (T_m) of the crystalline state of the material, if that exists. The glass transition temperatures (ranging between –70 and +215°C) of several polymers in routine use are shown in Table 5.1.

Consider the case of a rubber tire (styrene butadiene) that has –70°C as T_g. This material can exist in a rubbery elastic state down to –70°C, below which it solidifies. Thus, the increase in the brittleness of the rubber can be attributed to the decrease in temperature, as shown in Figure 5.5.

At $T > T_{trans}$, the original chain alignment of the polymer becomes deformed, dislocating the net points, and the material softens to accommodate the temporary shape with new chain–chain interactions. Thus, temporary shape training is complete and the polymer is cooled to $T < T_{trans}$, retaining the shape. Upon reheating beyond T_{trans}, the chain interactions formed during shape training are overcome, resulting in the material returning to its original shape. Thus, T_{trans}/T_g can be defined as the temperature beyond which the polymer softens for shape training, imparting *entropy elasticity*. Based on this interpretation, SMPs are copolymers with their hard segments acting as the fixed phase and soft segments acting as the reversible phase.

Chain mobility is a vital component in SMP behavior that in turn depends on the viscosity of monomers and the interactions between them. Chain mobility directly affects the transition temperature of the polymer, as will be explained in the following sections.

Using a blend of epoxy resin and cyanate ester–containing poly-(tetramethylene oxide) as the switching segment (T_g = 110°C), the permanent memorizable networks and reversible phases of SMPs have been studied to understand the basics of shape memory in polymers (Dyana Merline and Reghunathan

TABLE 5.1

Glass Transition Temperatures of Common Polymers

Sl No.	Polymeric Material	T_g (°C)
1	Styrene butadiene copolymer	–70
2	Polyvinyl fluoride (PVF)	–20
3	Polypropylene (PP isotactic)	0
4	Polyamide (PA)	47–60
5	Poly(vinyl chloride) (PVC)	80
6	Poly(methyl methacrylate) (PMMA atactic)	105
7	Poly(carbonate) (PC)	145
8	Polynorbornene	215

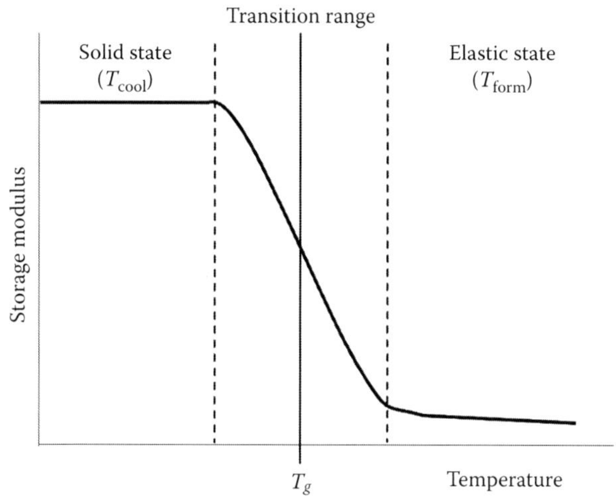

FIGURE 5.5
T_g expressed as a range of temperatures between the solid and elastic state; plot between storage modulus and temperature. T_{cool} is $T \ll T_g$ and T_{form} is $T \gg T_g$.

Nair, 2012). These studies have shown that the switching segment concentration decides the shape memory performance of the SMP system. The mechanical properties of the blend are significant, and an increase in creep temperature (the temperature range in which creep deformation may occur differs in various materials) shows higher deformation around T_g. The thermal transition in an SMP does not occur at a specific temperature but is a range over which the material transforms from a glassy phase to a rubbery state.

The range of the transition temperature is a function of the molecular mobility of the polymer chains in the SMP. A wider transition temperature range has a harmful impact on shape memory, ensuring unstable and partial shape fixing and slow recovery. In studies pertaining to the microstructures of SMPs, it has been observed that for both chemically and physically cross-linked SMPs, the shape-switching segments could be either crystalline or amorphous. The presence of SME can be ensured if the microstructure has mostly crystalline switching segments, even though amorphous switching segments with controlled glass transitions have been synthesized. The reaction conditions, curing time, cross-linking density, and viscosity of the monomers decide the shape memory as well as other properties.

As explained earlier, there are two types of cross-linking, physical (thermoplastics) and chemical (thermosets), of which chemically cross-linked SMPs show better mechanical properties. The polymer networks during a shape memory cycle in a polymer are illustrated in Figure 5.6.

For SME in polymers, a strong reversible bond/cross-link should be formed so that a temporary shape can be fixed (Figure 5.6b).

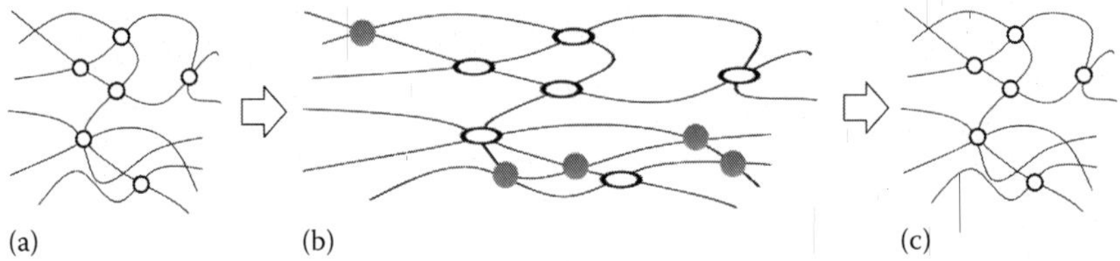

FIGURE 5.6
Polymer network configuration during a shape memory cycle. (a) Original shape. (b) Temporary/trained shape. (c) Recovered shape. (○ in the figure denotes the primary fixed cross-links and ● denotes the secondary reversible cross-links; the line segments denote polymer chains.) (Reprinted from Leng, J. et al., 2011, Shape memory polymers and their composites: Stimulus methods and applications, *Progress in Material Science*, 56(7): 1077–1135, with permission from Elsevier.)

Initially, the SMP's original configuration, as with any other polymer, is as it is depicted in Figure 5.6a. When heating the polymer to T_g and applying external stress, the segments/chains are stretched or elongated (Figure 5.6b), resulting in the alignment of the chains in the direction of the applied force. This naturally dislocates the physically or chemically cross-linked net points. The macroscopically deformed shape is maintained on cooling, forming secondary reversible cross-links. These secondary cross-linked domains can be attributed to the chain–chain interactions as they come closer to each other in the new alignment. Secondary cross-links can either be glassy or crystalline in nature. Reheating the polymer detaches the secondary cross-links, resulting in the release of the strain energy. SME is realized macroscopically and the original shape is recovered (Figure 5.6c).

Thus, SME is the result of the primary fixed cross-link domain, otherwise known as the *hard phase*, and the secondary reversible cross-link domain, or *soft phase*. During the shape memory cycle, the hard segments form net points and link the soft segments, which show thermal transition at the T_g of the polymer.

5.4 One-Way and Two-Way Shape Memory in Polymers

As with SMAs, SMPs also have multiple shape memory properties, such as one-way, two-way, and multiway shape memory. Once the permanent shape is recovered, a new shape-training step is required to rebuild the temporary shape in one-way SMPs (AB, after reprogramming AB in Figure 5.7). In two-way shape memory, the temporary shape changes upon terminating the stimulus and the permanent shape is regained (ABA, even when keeping the corresponding stimuli off, as in Figure 5.7).

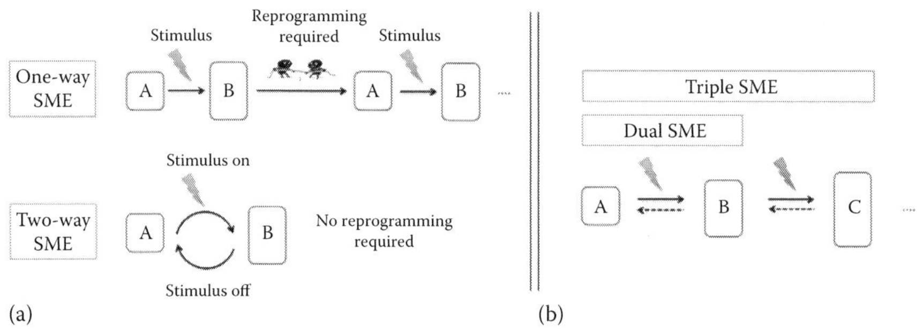

FIGURE 5.7
(a) One-way and two-way SME. (b) Dual and triple SME. (Reprinted from Hager, M. D. et al., 2015, Shape memory polymers: Past, present and future developments, *Progress in Polymer Science*, 49: 3–33, with permission from Elsevier.)

The multiple SMEs extending beyond one-way and two-way shape memory will correspond to the changes effected by the different transition temperatures, as mentioned previously, of the polymers that exist within one material.

SMPs are capable of fixing one or more temporary shapes (dual or multiple SME) and recovering the permanent shape in response to external stimuli. Referring to Figure 5.7, triple SMPs feature two temporary states (B and C) as well as a permanent shape (A). The polymer will go from shape C to shape A when induced by one trigger, whereas another trigger induces it to take back the original shape A from B. This concept can be extended to multiple shapes. When the transformation occurs with B alone as the temporary shape, it becomes dual shape memory, and upon extending the shape transitions between BA/CA, involving A, B, and C as three different shapes, it becomes triple shape memory.

There are two basic approaches to designing triple SMPs: (1) a very broad thermal transition temperature and (2) multiphase design (each phase provides a separate transition). In the first approach, the shape change occurs at two different temperature values within a single T_g (broad), whereas the second approach is achieved by two different points of transition/melting (Figure 5.7 illustrates one-, two-, and three-way SME). This is analogous to heating a mixture of two solid materials with two melting points. Even after melting one of the ingredients, the other one takes its own time to melt and become properly miscible.

Thus, if SMPs are able to memorize one temporary shape in each shape memory cycle, they are *dual SMPs* (BA in Figure 5.7). The ability of SMPs to memorize two temporary shapes (B and C) in one shape memory cycle is called *triple SME* (BA/CA in Figure 5.7). This second shape is memorized by creating a cross-link structure by adding a second reversible segment (responsible for shape C in this case), and both reversible segments should have a well-separated transition temperature (i.e., $T_gC \gg T_gB$). Hence, a well-separated T_g helps complete shape recovery in each case.

5.5 Comparison of SMPs and SMAs

SMPs can be easily synthesized or shape trained/programmed and can be tailor-made to specific applications. Compared with SMAs, SMPs are available at less cost and are more efficient, while many SMPs are biodegradable and have better mechanical properties than SMAs. A major difference is the nature of phase transition: one is glass transition while the other is martensitic transition (as explained in Chapter 2). For SMPs, the density ranges between 0.9 and 1.25 g/cc, while that of SMAs is between 6 and 8 g/cc. Another notable property difference is the wide range of transition temperatures of SMPs compared with that of SMAs. This wider thermal transition zone is the reason for unstable, partial shape fixing and slow recovery in SMPs, as discussed earlier. The range of an SMP's thermal transition (Figure 5.5) depends on the distribution of relaxation time associated with the molecular mobility of the polymer chains, which in turn is dictated by the chemical composition and network structure of the SMP.

A well-separated Young's modulus on either side of T_g is a specialty of SMPs. The stress levels of SMPs (1–3 MPa) are much lower compared with SMAs, which require 50–200 MPa. Due to the inherent properties of polymers, the thermal/electrical properties of SMPs are inferior to SMAs. The poor thermal/electrical properties of SMPs are regarded to be the bottlenecks when compared to SMAs. The improvement in those properties in SMPs by the addition of fillers (fillers can act as carriers of electrical charge or as local conductors in polymer matrices) is being explored by many researchers and scientists throughout the world. The fabrication of SMPs as well as their processing conditions and shape-training processes are less tedious compared with those of SMAs, which is one of the operational advantages of SMPs over SMAs. SMAs are preferred over SMPs for most applications due to their higher actuation forces and faster responses. The challenges to overcome this are being studied for SMPs by improving the response properties of the SMP matrix, as well as by embedding fillers in a composite material level.

SMPs qualify for highly demanding applications due to their 99% recovery precision. SME in SMPs is a function of their components, whereas for SMAs, the alloy composition dictates SME. For polymers, the weight fraction of the reversible segments and the molecular weight of the polymer chain play major roles in SME.

Some uses of SMPs as basic ingredients in composites are deployable space structures, self-healing materials, passively deployed solar arrays, textiles, cardiovascular stents, biocompatible microscale neuronal probes, and other biomedical applications.

To bridge the gap between synthesis, processing, structure, and morphology at the micro scale and material and device performance at the macro scale, the molecular mechanisms of SMP behavior must be understood. At the structural level, the concern is mainly about the shape memory recovery rate, the stress generated in the material during temperature cycling, the modulus ratio between low- and high-temperature states, the retention/loss of SMP behavior, and aging.

As discussed earlier, shape recovery or phase transformation is initiated by some form of energy that can cause microscopic rearrangement reflected at the macroscopic level. This energy injection into the system stimulates the shape change and has a significant role in the various mechanical properties, recovery time, and stability.

5.6 Response Stimuli

Shape recovery in a shape memory material is achieved by releasing the strain energy stored in the temporary shape by heating it to T_g or by exposing it to a specific stimulus, as with SMAs. The most common and least complex method of triggering used so far is thermal stimulation or simply heating the element to return it to its original shape. This heating can be effected on the system in various ways by direct or indirect means. Direct heating is provided by an external flame or a medium that can transfer heat to the element under observation. Indirect heating can be achieved with electrical energy through Joule heating, induction and mechanical heating, laser light sources, and so on. Almost all SMPs can be categorized into three major classes based on the stimulus applied to induce the SME: thermoresponsive SMPs, photoresponsive SMPs (i.e., those affected by different wavelengths/frequencies of light, regardless of heat), and chemoresponsive SMPs (i.e., those affected by chemicals).

In many cases, exposure to light results in heating and thus leads to thermal transitions. These materials can be classified as light-activated SMPs (LASMPs). The most prominent example is the reversible photodimerization of cinnamic acid derivatives. The two main stimuli applied to SMPs are temperature and, to a lesser extent, light. One experiment on photosensitive SMP materials revealed that star-poly(ethylene glycol) end-capped with cinnamylidene acetic acid terminal groups (SCAA) yields an interpenetrating photosensitive network that changes shape in response to light of a particular wavelength. The photosensitive groups in SMPs act as molecular switches in light-activated memory materials. LASMPs are basically realized by incorporating reversible photoreactive molecular switches that react when a unique wave number/frequency of light falls on them. This is free from any form of heating in its stimulation or actuation stages.

However, further stimuli can be utilized to trigger a transition (e.g., segmental rearrangement) that induces SME. Cross-linked natural rubber can also be triggered by organic solvents that induce rearrangement within the polymer.

Interestingly, high-intensity focused ultrasound has also been applied to trigger SME in polyurethanes. During ultrasound treatment, the temperature within the polymer increases, which consequently leads to SME.

A few polymers exhibiting SME upon contact with chemicals and water have also been discovered. Chemoresponsive SMPs are of particular interest as the bioclinical applications of SMPs can be better effected by this stimulus. A depiction of a water-induced SMP is shown in Figure 5.8.

In Figure 5.8, a polyurethane wire with a particular T_g value is made into a closed loop and introduced to water (0 min). Once the water particles have begun to diffuse it at 30 min, the loop is untied slowly through 40, 60, 80, 100, 120, and 140 min, eventually achieving its original open configuration. This is achieved not by heating the material to T_g but by lowering the T_g value to the temperature of water and thus realizing SME, which is the specialty of water-driven shape memory.

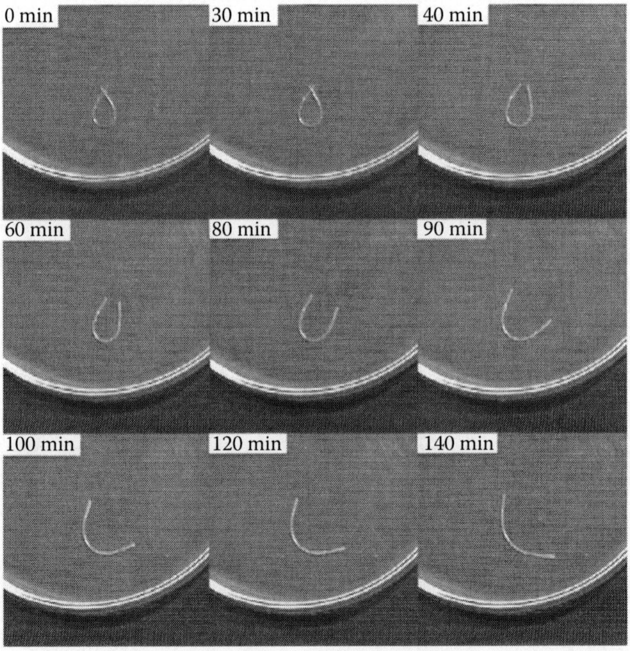

FIGURE 5.8
Shape memory polyurethane water-driven actuation stages. (Reprinted from Leng, J. et al., 2011, Shape memory polymers and their composites: Stimulus methods and applications, *Progress in Material Science*, 56(7): 1077–1135, with permission from Elsevier.)

For water-triggered SME, as the material is immersed in water, the solvent molecules diffuse into the polymer sample and impart softness and flexibility to the system. These softer and more flexible materials, known as *plasticizers*, cause the flexibility of polymer chains (the overall softening of the materials) and thus lead to the reduction of the transition temperature, resulting in shape recovery. In this case, instead of heating the material to above its T_g, shape recovery can be achieved by means of reducing the T_g of the material.

The electric stimulation of SMPs is one of the most researched areas as it opens horizons to remote controllability, which is significant in the case of aerospace applications.

Electrically nonconductive SMPs are made conductive by adding filler materials such as carbon-based particles, tubes, fibers, or planar sheets (nanopaper or graphene) in an optimal quantity resulting in carbon composites. Thus, the electric current passed through them is converted into heat, and shape training, fixing, and recovery are initiated in electroactive SMPs.

Magnetism is another triggering mechanism that facilitates the noncontact mode of shape changes in polymers, and this is achieved by inductive Joule heating in alternating magnetic fields. Thermal and magnetic dual-stimulus SMPs are realized by incorporating magnetite nanoparticles into a grafted polymer network of poly(ethylene glycol) side-chains and poly(caprolactone) cross-links. Noncontact triggering mechanisms are considered the future of SME systems and have been demonstrated by magnetically inducing SME in composites. This is achieved by incorporating magnetic nanoparticles and applying inductive heating in an alternating magnetic field (Mohr et al. used frequency $f=258$ kHz and magnetization $H=30$ kAm^{-1} to verify the concept). The noncontact triggering of SMP composites has promising applications in the medical field, as smart implants and surgical instruments could enable medical practitioners to perform mechanical adjustments on a noncontact basis.

Stimulation with thermal energy from infrared light is another possible method of indirectly thermally inducing SME in polymers. These kinds of SMPs are supplemented with conducting fillers such as CB, to which resistive heating is applied during illumination with the radiant thermal energy of infrared light. CB absorbs infrared radiation (IR), resulting in a heating effect for IR wave numbers between 500 and 3000 cm. The processes of shape recovery during heating are stress-free recovery cycles and constant–strain recovery cycles (fixed deformations are held when the material is heated above T_g).

5.7 Thermoset and Thermoplastic SMPs

There are two types of polymers based on cross-linking: physically cross-linked polymers are termed *thermoplastics* and those chemically cross-linked are *thermosets*. The primary physical difference is that thermoplastics can be remelted back into a liquid, whereas thermoset plastics remain in a permanent solid state.

Thermosets usually possess a higher T_g than the corresponding thermoplastic polymers, and the increase of T_g for thermoset SMPs is a result of the disappearance of chain ends and the formation of chemical cross-links. The T_g of thermoset samples ranges from 240°C to 250°C, which is suited to environments in space. Thermoset SMPs are expected to have applications in many harsh environments, as they are intrinsically chemical resistant and possess higher tensile strength than their thermoplastic counterparts. Thermoset SMP polyimides have been found to have shape fixity and recovery at nearly 100%. Thus, by controlling their cross-link densities and the rigidity of the polymer backbones, SMPs with tunable T_g, mechanical properties, and better SME can be synthesized for specific applications.

In thermoplastic SMPs, the chain entanglements or local crystalline domains formed by strong chain–chain interactions can serve as the physical cross-linking sites. The advantages of thermoplastic SMPs include their moldability into different permanent shapes/configurations, wide shape deformation range, and ease of blending with other polymers and additives.

It has been verified that the T_g of polyimide polymer shape memory material can be adjusted by controlling the molecular weight of thermoplastic SMPs (depending on the cross-link of the polymer chains). Thermoset shape memory polyimides show a higher T_g and better shape memory performance than thermoplastics due to the disappearance of chain ends and the formation of chemical cross-links.

Figure 5.9 is a schematic of how shape training and recovery occurs in thermoset and thermoplastic SMPs with physical cross-links (Figure 5.9a) and those with covalent cross-links (Figure 5.9b). Chemically cross-linked shape memory can be imagined as a fishnet deformation, which, due to the presence of continuous networks/bonds, gains back its shape upon the release of stress.

T_g is influenced by the modulus, which is influenced by the cross-link (the stiffness variation of SMPs is effected by cross-linking), which in turn is a function of the viscosity of the polymer, and viscosity is a component of the molecular weight of the polymer. The relationship between the T_g and molecular weight (M_n) of the shape memory polyimide can be determined as follows:

$$T_g = T_{g\infty} - \left(K_g / M_n\right) \tag{5.3}$$

where:

$T_{g\infty}$ is the T_g of the sample with infinite molecular weight

K_g is the parameter explaining the M_n dependence of T_g for the material

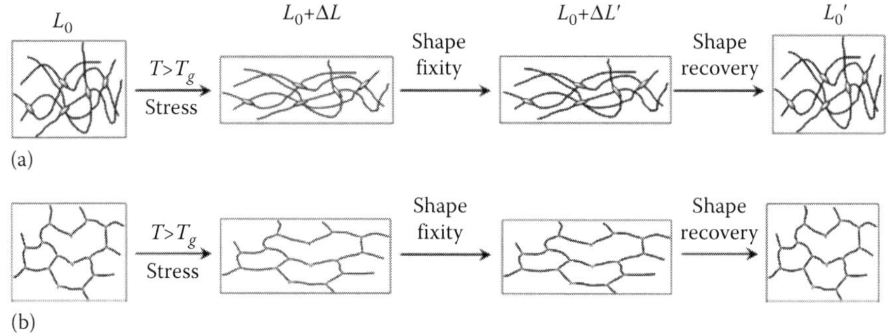

FIGURE 5.9
Shape memory behavior of polyimides. (a) Thermoplastic polyimides with chain entanglements. (b) Thermoset polyimides with covalent cross-links. (Adapted from Xiao, X. et al., 2015, Shape-memory polymers with adjustable high glass transition temperatures, *Macromolecules*, 48(11): 3582–3589, with permission from the American Chemical Society.)

5.8 Cross-Linking

Chemically cross-linked SMPs generally exhibit better strain-fixing and strain recovery ratios, faster strain recovery rates and greater shape recovery stress, and most importantly, lower strain-to-failure values. Such a combination may arise from their higher cross-linking densities and the ability of the chemical cross-links to withstand tensile stress.

The main disadvantages of thermoset SMPs are the need for tailored polymer processing and their inability to be reprocessed into a new shape/configuration after being chemically cross-linked, which may not be a concern when the SMP is designed for one-time applications. Thus, an ideal SMP system may lie at the interface of thermoplastics and thermosets, as understood from the previous discussions.

Physical cross-links have high strain recovery, whereas chemical cross-links have high stress recovery. Many challenging applications require high failure strain and good mechanical stability. Shape memory epoxy resins (SMEPs) (thermosets with 100% shape recovery and good shape fixity as experimentally shown; T_g between 60°C and 150°C) generally have good mechanical attributes but poor elongation. By using aromatic/aliphatic epoxy resin in combination with aliphatic diamine as a curing agent, it is possible to achieve a high-strain and low-rubbery modulus. The inclusion of nanomaterials will bridge the lacking properties of epoxy resins with shape memory behavior (a discussion of which is detailed in Section 7.4). SMEPs fall into the shape memory polymer composite (SMPC) category and are good enough to replace SMAs, as they have both shape memory properties and structural stability with the added advantage of the low density of polymers. Studies show that, for SMEPs, increases in cross-link density increase the recovery speed. They are suitable for processing good elastic memory composites, as cyanate esters normally promote excellent composite formation, which have controllable transition temperatures and good processability.

The *storage modulus* of a polymer measures the stored energy, representing the elastic portion, and the *loss modulus* measures the energy dissipated as heat, representing the viscous portion. These are the key factors that help to determine the elastic modulus of an SMP, as explained in the following section.

When loading a viscoelastic material, the applied external energy is split into two parts by the material, out of which one part is stored for recovery (due to its elastic properties) and the other part is dissipated as heat or any other form of energy (due to its viscous properties). The storage modulus measures the stored energy and the loss modulus measures the energy dissipated.

The high-storage modulus in the glassy state of polymers originates from its potential elasticity, and the low-storage modulus in the rubbery state originates from the entropy elasticity due to micro-Brownian movement (the vibration or movement of microparticles due to the collision of fluid particles at high energy levels).

The thermodynamic explanation of SME in polymers is given by Boltzmann's entropy formula, which refers to polymer chains that are randomly distributed and coiled within the matrix in an amorphous state.

Boltzmann's entropy formula gives the relationship between the entropy of the system and the probability of a coiled conformation.

Mathematically, entropy is expressed as

$$S = \mathrm{k} \ln W \tag{5.4}$$

where:

k is Boltzmann's constant

W is the probability of a strongly coiled conformation (conformation with maximum entropy)

Thus, entropy is defined as measure of the number of possible microscopic states (or microstates, expressed as maximum entropy conformation) of a system in thermodynamic equilibrium, consistent with its macroscopic thermodynamic properties (accommodated by Boltzmann's constant).

When a polymer is in the transition from a glassy state to an elastic state by thermal activation, the rotations around segment bonds become unrestrained. Thus, the polymer chains are untangled to create new conformations that result in the densification of the microstructure. Therefore, the value of W in Equation 5.4 increases. Upon the application of an external force to train the shape, there is a minimum critical duration below which the strained coils will return after the external stress is removed, due to the

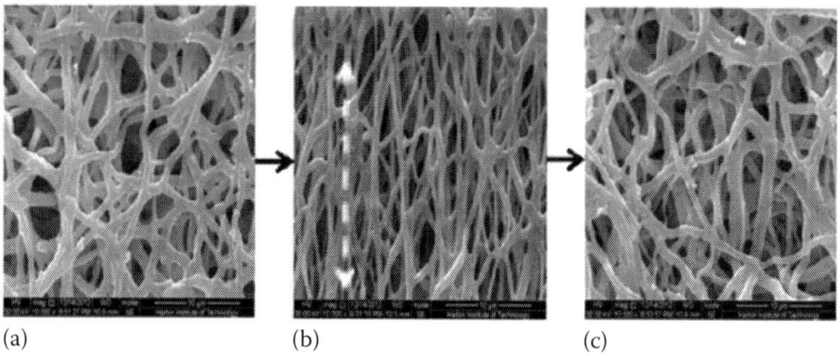

FIGURE 5.10
SEM images of SME in Nafion nanofibers. (a) Original fibers. (b) Stretched fibers. (c) Recovered fibers. (From Zhang, F. et al., 2015, Shape memory polymer nanofibers and their composites: Electrospinning, structure, performance, and applications, *Frontiers in Materials*, 2: 1–10.)

neighboring chain interactions. If the applied stress is extended beyond the time limit, then the chains will undergo permanent decoiling/slipping. This can be prevented by cross-linking the polymer (chemically or physically).

Nanoscale fiber stretching in shape memory has been reported in the literature (Zhang et al., 2015), where the original fiber is stretched and recovered during a cycle, as depicted in Figure 5.10a–c.

The relationship between energy (e.g., deformation and stored strain energies) and recovery stress is highly advantageous since it could be used to predict the ability of SMPs to perform under partially constrained conditions (e.g., the stretching and recovery of fibers shown in Figure 5.10a–c). Energy loss due to thermal transition could be quite significant in this case. Recovery stress relates to the energy stored during deformation, and SMPs are deformed in the rubbery state, thus limiting the energy stored, resulting in relatively low shape recovery stress compared with SMAs.

A structural characterization of thermally stimulated (direct or indirect) SMPs can be summarized as follows in line with the preceding discussions; this will give researchers a choice of SMP based on their specific requirements. The following sections cover physical and chemical cross-links as detailed previously, and thermoplastics are subdivided into branched and linear, with suitable examples for each. Linear SMPs are further subcategorized into block copolymers and those with high molecular mass. Similarly, examples of thermosetting with a wide domain are listed in Figure 5.11.

Traditional SMPs can be classified into four categories, based on the nature of the thermoreversible transition and the cross-linking:

1. Chemically cross-linked glassy polymers
2. Chemically cross-linked semicrystalline polymers
3. Physically cross-linked glassy polymers
4. Physically cross-linked semicrystalline polymers

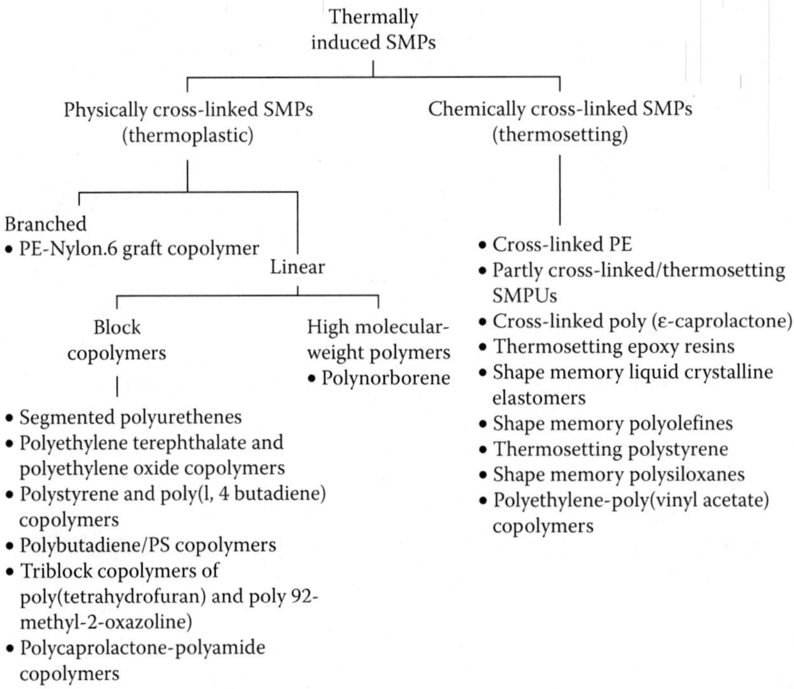

FIGURE 5.11
Structural categorization of SMPs. (Reprinted from Leng, J. et al., 2011, Shape memory polymers and their composites: Stimulus methods and applications, progress in materials science, *Progress in Material Science*, 56(7): 1077–1135, with permission from Elsevier.)

As per the polymer classification convention, the first two categories belong to thermosets and the last two to thermoplastics. These classifications are based on differences in fixing mechanisms and the origin of "permanent" shape elasticity (Figure 5.12).

5.8.1 Covalently Cross-Linked Glassy Thermosets

The simplest type of SMP is a cross-linked glassy polymer featuring a sharp T_g at the temperature of interest and rubbery elasticity above T_g derived from covalent cross-links (Figure 5.12a). These classes of materials have attractive characteristics, including an excellent degree of shape recovery. This is due to the nature of permanent (or near-permanent) cross-linking, tunable workability due to a rubbery modulus that can be adjusted through the extent of covalent cross-linking, and an absence of molecular slippage between chains due to strong chemical cross-linking.

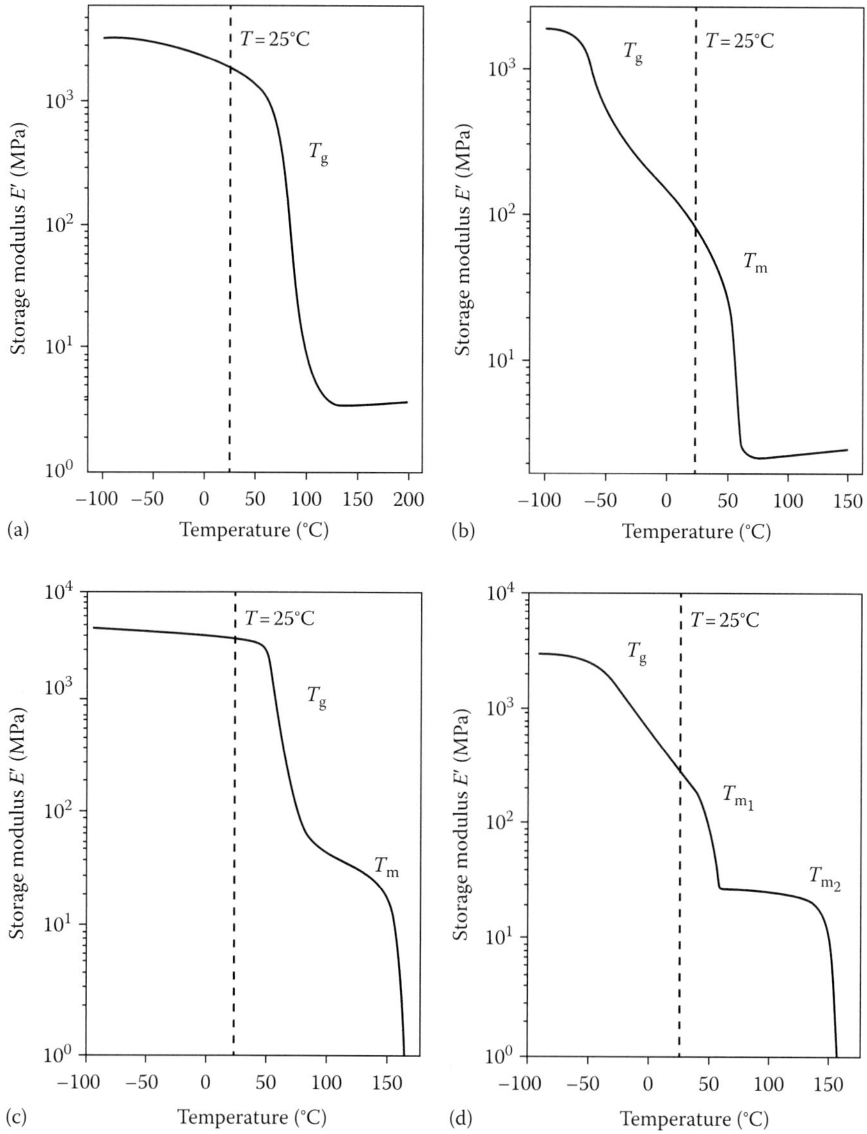

FIGURE 5.12
Tensile storage modulus vs. temperature as measured using a small oscillatory deformation (1 Hz). (a) Chemically cross-linked glassy thermosets. (b) Chemically cross-linked semicrystalline rubbers. (c) Physically cross-linked thermoplastics. (d) Physically cross-linked block copolymers. (Adapted from Liu, C. et al., 2007, Review of progress in shape-memory polymers, *Journal of Materials Chemistry*, 17(16): 1543–1558, with permission of the Royal Society of Chemistry.)

5.8.2 Covalently Cross-Linked Semicrystalline Thermosets

Aside from glass transition at a critical temperature, the melting transition of semicrystalline networks can also be employed to trigger shape recovery, typically giving a sharper recovery event (Figure 5.12b). Here, the secondary shapes are fixed by crystallization instead of being fixed by vitrification. Similar to the first class, the permanent shapes are established by chemical cross-linking and cannot be reshaped after processing.

5.8.3 Physically Cross-Linked Glassy Copolymers

In this SMP class, crystalline or rigid amorphous domains in thermoplastics may serve as physical cross-links, affording the super-T_g elasticity required for shape memory to be developed, mainly in the form of phase-separated block copolymers (Figure 5.12c). When the temperature surpasses the T_m or T_g of these discrete physical domains, the material will flow and can be processed and reshaped.

5.8.4 Physically Cross-Linked Semicrystalline Block Copolymers

For some block copolymers, the soft domain will crystallize and, instead of T_g, their T_m values function as the shape memory transition temperatures; the secondary shapes are thus fixed by the crystallization of the soft domains Figure 5.12d).

The chemical cross-linking of polymeric binders is possibly the most significant factor affecting the mechanical behavior of unfilled and highly filled elastomers. There are basically two physical methods used by various investigators to experimentally determine the degree of cross-linking in elastomeric materials. They are the swelling method using Flory's equation and the equilibrium stress–strain method.

$$F/A_i = \nu e \times RT\left(\lambda - \lambda^{-2}\right) \tag{5.5}$$

where:

- F is the force, g/cm^2
- A_i is the initial cross-section area of unstrained specimen, cm^2
- νe is the volume fraction
- R is the universal gas constant (8.479e4 J/mol K)
- T is the test temperature, K
- λ is the extension ratio

The Flory–Rehner equation explains the mixing of polymer and liquid molecules as predicted by the equilibrium swelling theory of Flory and Rehner. It describes the equilibrium swelling of a lightly cross-linked polymer in terms of cross-link density and the quality of the solvent.

The theory is based on the following considerations:

- The entropy change caused by mixing polymer and solvent
- The entropy change caused by the reduction in numbers of possible chain conformations via swelling
- The heat of mixing of polymer and solvent, which may be positive, negative, or zero

The swelling method shows that the equilibrium swelling of the polymer in a good solvent is obtained (allowing the polymer to swell for 1–2 weeks in the dark to reach equilibrium). The swollen gel is then isolated and weighed by vacuum drying.

The cross-link density (U) according to the Flory–Rehner equation (network chain per gram) is

$$U = -\left(\ln\left(1-V_p\right)+V_p+\left(V_p\right)^2 \times X\right)\Big/\left(D_p\left(V_s\right)\left(\left(V_p\right)^{(1/3)}-0.5V_p\right)\right) \quad (5.6)$$

where:

V_p is the volume fraction of polymer in swollen polymer
V_s is the molar volume of solvent (cm^3/mol)
X is the Huggins polymer–solvent interaction
D_p is the density of polymer (g/cm^3)
$V_p = 1/(1+Q)$

where Q = (weight of solvent in the swollen polymer × D_p)/(weight of polymer × D_o) and D_o is the density of solvent (g/cm^3)

Cross-linking the thermoset SMP is the key to the rubbery-state elastic modulus, and experimental results suggest that a suitable filler material can improve this property. Silicon carbide particles, CB, carbon nanotubes (CNTs), glass fibers, and aramid fibers have been experimentally tested to improve the elastic modulus; however, a decreased shape recovery percentage was observed. The electrical conductivity increased, enabling potential electrical field stimulation by the addition of carbon-based nanoparticles, nanotubes, and nanofiber fillers. Studies have observed that increasing the cross-linking density results in a higher T_g and, therefore, higher switching temperatures, faster shape recovery, better mechanical properties, and better shape memory properties (high recovery/fixity), which helps to design SMPs with predetermined properties.

5.9 Wax Analogy of SMPs

SME in polymers can be modeled and illustrated using the analogy of paraffin wax (Figure 5.13). By heating wax until it partially melts, the solid part acts as the elastic component, while the melted part functions essentially as the switching segment. An indentation is made at 70°C in the middle of the

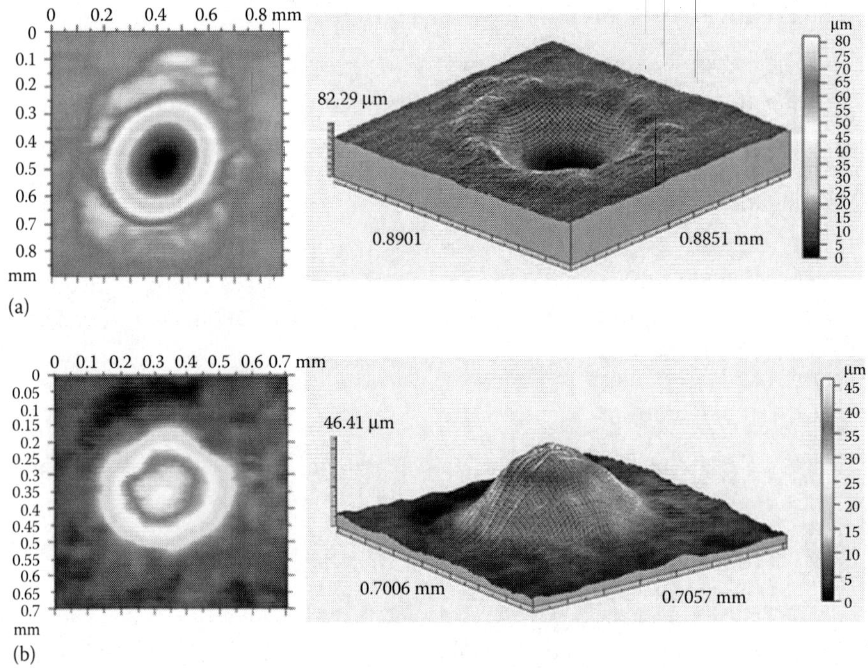

FIGURE 5.13
Wax analogy of SME. (a) Surface morphology after intendation (deformation). (b) Surface morphology after shape recovery. (Reprinted from Huang, W. M. et al., 2012, Thermo/chemo-responsive SME in polymers: A sketch of working mechanisms, fundamentals and optimization, *Journal of Polymer Research*, 19(9): 952–986, with permission from Elsevier.)

melting area under observation and then it is cooled to room temperature, after which the top surface of the specimen is polished to remove the indent. Upon subjecting it again to heating at 70°C, a protrusion is observed, which reflects the behavior of a shape memory material.

5.10 Molecular Mechanisms in SMPs

Shape-switching components are the polymeric chains in an SMP network that can switch from one state to another in response to temperature changes and are responsible for temperature-dependent deformation and recovery. Thus, the shape-deforming temperature (T_d) is the working temperature at which the SMP is transformed to a temporary shape (i.e., the shape-training stage). The relationship of T_d to T_g (i.e., whether it is below, at, or above it) has a significant impact on the shape memory performance of an SMP.

The temporary shape-fixing temperature (T_f) is the working temperature at which the temporary shape of a deformed SMP is fixed. T_f is usually lower than T_g. The shape recovery temperature (T_r) is the working temperature at which an SMP is triggered to recover from its fixed temporary shape. T_r is usually higher than T_g and is often chosen to be the same as T_d.

The morphology of SMPs, as previously discussed, is the combination of a switching phase and an elastomeric phase. The amount of miscibility between them (the ratio between the hard and soft segments) determines the viscoelastic properties, which in turn influence the modulus and thus affect the shape memory behavior. Thermoset shape memory polyimides exhibit better shape memory performances than their thermoplastic counterparts, mainly due to the loss in the physical cross-link integrity of thermoplastic polyimide caused by mechanical deformation.

The morphological features of SMP blends of styrene–butadiene–styrene (SBS) triblock copolymer and poly(3-caprolactone) (PCL) have been studied to understand the general rules of SMP morphology, and the following observations were recorded (Figure 5.14a–d). A combination of both components in wt % varying from 0 to 100 was used for the investigation. The T_g of the blend was between 100°C and 120°C. Atomic force microscopy (AFM) of each combination reveals the internal structure, which is depicted in Figure 5.14.

The elastomeric (thermoplastic) and switch polymer (any crystalline polymer) were found to contribute to the shape recovery and fixing performances, respectively.

By carefully designing the immiscible phase morphology, an ideal SMP system with both good stability and performance can be achieved; Figure 5.14b with elastomer as the major continuous phase and the switch polymer as the minor phase could be concluded as the optimized combination in a shape memory context. In this optimized design, the elastic matrix provides good stretching and recovery performance, while the continuous switch polymer phase provides good fixing and unfixing performance while hardly reducing the recovery performance. A general shape memory mechanism for this type of polymer-blend SMP has been proposed in which the two immiscible components of the blend separately contribute to the shape memory performance. The elastomer provides the stretching and recovery performances and the switch polymer provides the fixing and unfixing performances. The phase morphology is thus optimized, which contains the elastomer's major continuous phase and switch polymer's minor continuous phase.

The phase that fixes the temporary shape in an SMP can also be based on reversible bonds or covalent interactions. Supramolecular interactions, rather than reversible covalent bonds, are utilized in the design of a reversible network, and the possible methods are hydrogen bonding, ionic interactions, or metal ligand interaction.

The morphology of phase separation between SBS and PCL in their blends is induced by the immiscibility of the three components. With the increasing PCL content, the PCL phase transforms from a droplet-like dispersion phase to a continuous phase, and then to the matrix, as in Figure 5.14.

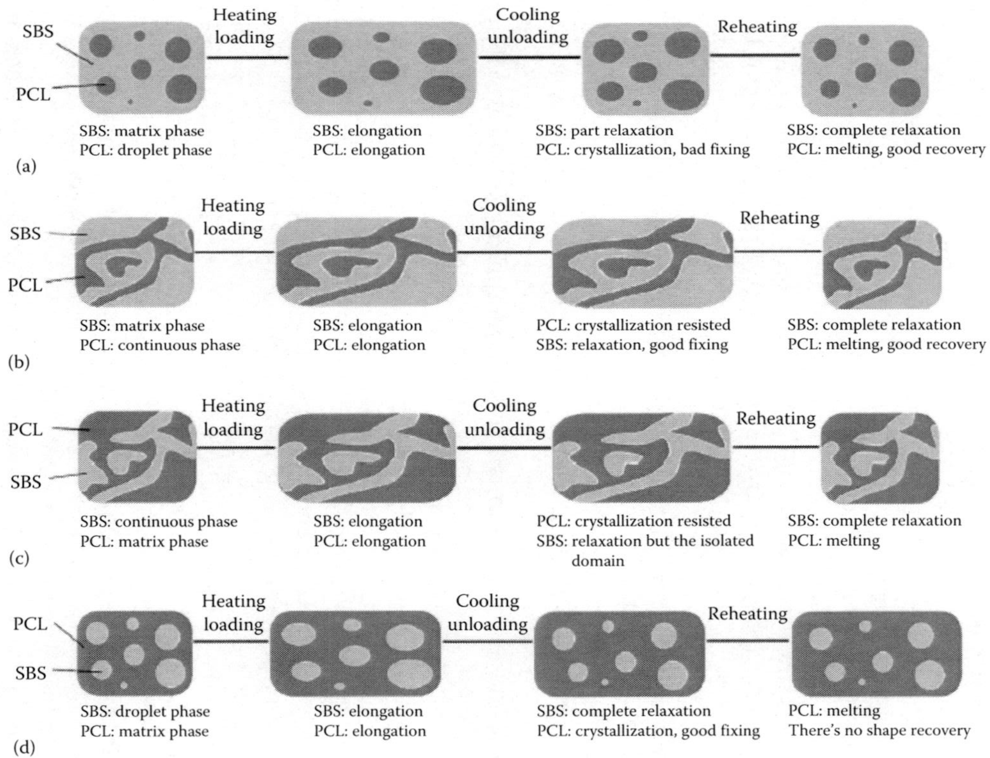

FIGURE 5.14
Schematic of the shape memory mechanism of elastomer/switch polymer blends concluded from SBS/PCL blend. (Reprinted from Zhang, H. et al., 2009, A novel type of shape memory polymer blend and the shape memory mechanism, *Polymer*, 50(/6): 1596–1601, with permission from Elsevier.)

5.11 Self-Healing SMPs

Any material capable of repairing mechanical damage by itself is defined as a *self-healing material*; this is one of the most promising fields of application for SMPs. The underlying principle is the activation of polymer chain networks via thermal stimulation, which causes the embedded linear polymer chains to diffuse and heal damaged areas.

Two different methods of obtaining self-healing polymers are by embedding healing agents (e.g., capsules or vascular systems) and introducing reversible interactions (e.g., reversible covalent bonds). The first method (Figure 5.15a) involves packages of healing compounds that sense and act

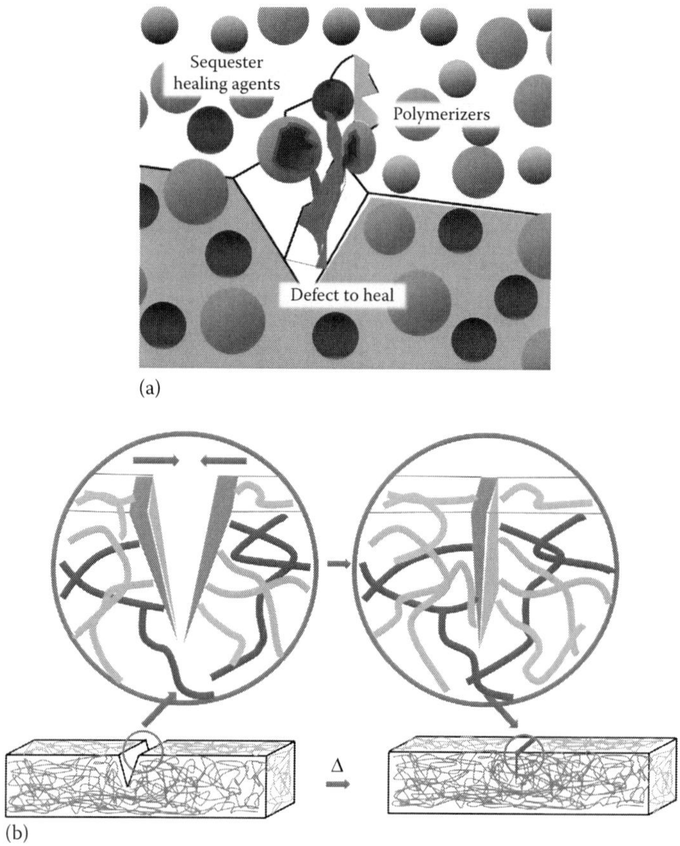

FIGURE 5.15
(a) Microcapsules embedded inside the smart material break and fill the crack to heal it. (From Sinn-Hanlon, J. et al., 2012, https://sites.google.com/site/shmolfw/design-strategies/release-of-healing-agents.) (b) At the glass transition temperature, the chains become more mobile and allow the materials to flow into each other, forming new bonds. (Reprinted from Luo, X., Mather, P. T., 2013, with permission from the American Chemical Society.)

based on the situation, while the second (Figure 5.15b) is similar to the rejoining of bones in the human body (the rebonding of covalent interactions as the reactive sites reopen).

A major unaddressed issue in this regard is the requirement that the shape memory part must be located in a position close to the crack/damage, otherwise SME may not have any use in the self-healing process. The *bulk shape memory assisted self-healing* (SMASH) system was developed to address this issue, whereby the aforementioned packages are distributed evenly to act in the proximity of a crack and heal it.

5.12 Characterization and Testing

The material in use should have a known structure and predictable properties and behaviors in order to suit it to different applications/experiments. The different methods of probing into SMPs to ascertain their properties are explained in this section. These methods include thermal characterization (differential scanning calorimetry [DSC], thermomechanical analysis [TMA], dynamic mechanical analysis [DMA]), cross-link density (explained in Section 5.5.1), SMP morphology, cyclic/bending tests, elastic modulus, and theoretical approaches (modeling).

As discussed earlier, T_g is one of the most important properties of any polymer as it is the temperature region in which the polymer transitions from a hard, glassy material to a soft, rubbery material. T_g is not a discrete thermodynamic transition point but a temperature range/zone over which the mobility of the polymer chains increase significantly. The factors associated with T_g are

- The chemical structure of the matrix resin
- The type of hardener
- The degree of cure

T_g strongly depends on the cure schedule, and thus a low-temperature cure (room temperature) will result in the lowest possible T_g, and if the same material is cured at an elevated temperature, a higher T_g will result.

The methods to determine the T_g of an SMP are briefly explained in the following sections.

5.12.1 Differential Scanning Calorimetry (DSC)

This is a thermoanalytical technique in which the difference in the amount of heat required to increase the temperature of a sample and a reference is measured as a function of temperature. Both the sample and reference are maintained at roughly the same temperature throughout the experiment.

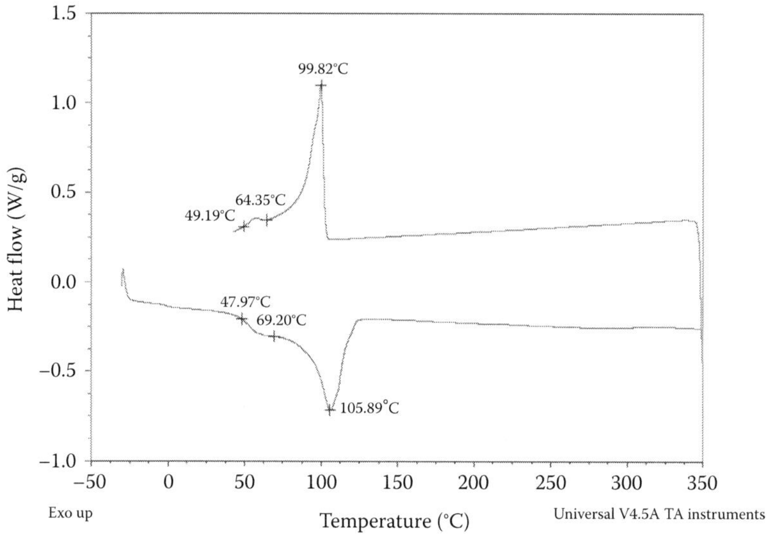

FIGURE 5.16
DSC plot for polyethylene.

Generally, the temperature program for a DSC analysis is designed such that the sample holder temperature increases linearly as a function of time. The reference sample should have a well-defined heat capacity over the range of temperatures to be scanned. The temperatures are recorded during the heating and cooling stretch of the cycle.

This technique was developed by E. S. Watson and M. J. O'Neill in 1962 and introduced commercially at the 1963 Pittsburgh Conference on Analytical Chemistry and Applied Spectroscopy. The first adiabatic differential scanning calorimeter used in biochemistry was developed by P. L. Privalov and D. R. Monaselidze in 1964. The term DSC was coined to describe an instrument that measures energy directly and allows precise measurements of heat capacity.

A polyethylene sample was subjected to the heating and cooling cycles of the DSC and a plot was obtained, as depicted in Figure 5.16. The T_g of the sample in the cool cycle is between 49.19°C and 64.35°C, while the same in the heat cycle is between 47.97°C and 69.20°C. The melting points (T_m) in the cooling and heating cycles are 99.82°C and 105.89°C, respectively.

DSC is a cost-effective method; however, it is not always accurate, as many materials will only show a small change in heat capacity, making the transition difficult to identify.

5.12.2 Thermomechanical Analysis (TMA)

This method determines the T_g of an SMP/polymer by tracking its coefficient of thermal expansion (CTE). The CTE of a polymer material increases three to five times during its transition from the glassy to the rubbery state.

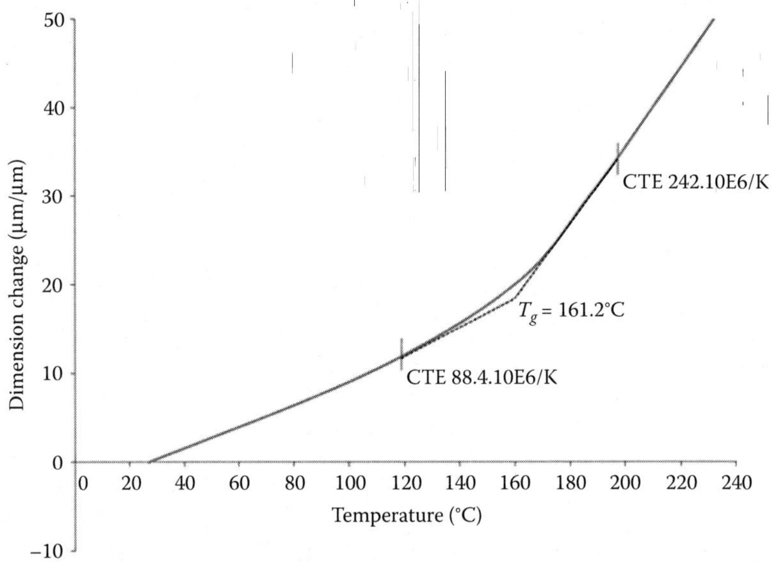

FIGURE 5.17
TMA curve for a fiberglass polyester composite prepreg material.

The curve in Figure 5.17 shows the T_g and the difference between the CTE below and above the T_g of the polyester component of the composite.

Similar to DSC, TMA also involves the slow addition of heat to the sample while the changes in physical dimensions are being observed. The physical dimensions of the material are tracked, and the T_g is identified as the temperature at which there is a dramatic change in these dimensions. TMA yields better accuracy compared with DSC.

5.12.3 Dynamic Mechanical Analysis (DMA)

DMA determines the mechanical properties of a material by applying a sinusoidal load to a specimen and measuring the resultant deformation by subjecting the sample to temperature changes. This reveals the mechanical (storage and loss moduli) and thermal (T_g) properties of polymeric materials over a wide range of temperatures (–150°C to 800°C). Figure 5.18 shows the DMA plot of poly(ε-caprolactone) (PCL3000AT/4000AT), which shows the melting transition T_m sharply decreasing in the storage modulus when heated. The sudden decrease of the storage modulus value is also observed at the transition temperature.

In the category of elastomers, thermoset and thermoplastic SMPs undergo melting transition and glass transition, respectively. Shape recovery by melting the switching segments is utilized in chemically cross-linked semicrystalline polymeric networks and physically cross-linked polymers. Shape recovery using T_g as the transition temperature is utilized

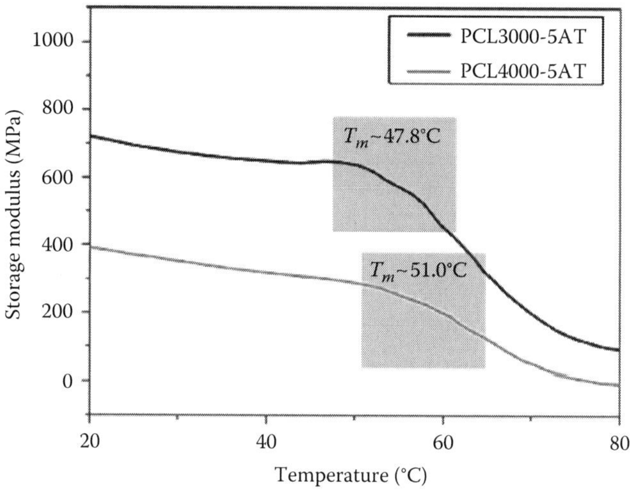

FIGURE 5.18
DMA plot of electroactive copolymers. PCL*x*-*y*AT means that the molecular weight of PCL is *x* and the weight percentage of aniline trimer (AT) is *y*% in the copolymers. (Reprinted from Deng, Z. et al., 2016, Stretchable degradable and electroactive shape memory copolymers with tunable recovery temperature enhance myogenic differentiation, *Acta Biomaterialia*, 46: 234–244, with permission from Elsevier.)

in chemically cross-linked thermosets as well as physically cross-linked thermoplastics. SMPs based on polyolefins, polyethers, and polyesters are T_m-type materials that typically feature a glass transition far below room temperature, preventing the application of T_g as the switch temperature for SME. Thus, polymeric materials with T_g above room temperature can be utilized for T_g-based SMPs. Experiments show that, in comparison with T_m-based SMPs, T_g-based SMPs are slow in shape recovery due to their broad glass transition temperature ranges, making them not suitable for applications requiring sudden shape recovery but acceptable for those in the biomedical field.

5.12.4 Cyclic/Bending Tests

Cyclic or shape memory bending tests are performed to judge the recovery ratio/shape fixity/efficiency of the SMP in real-world applications. A typical bending test with a predesigned mold is depicted in Figure 5.19. In this case, a new thermosetting shape memory cyanate polymer (SMCP) is modified with polybutadiene/acrylonitrile (PBAN) and proves to be a competent high-temperature SMP for aerospace applications (T_g around 255°C). It shows a T_g of 241.3°C and is very tough, making it suitable f™or environments in space.

A standard U-shaped mold and a fitting mandrel deform the SMP specimen from its original shape at $T_g + 40°C$, as shown in Figure 5.19. It is subjected to

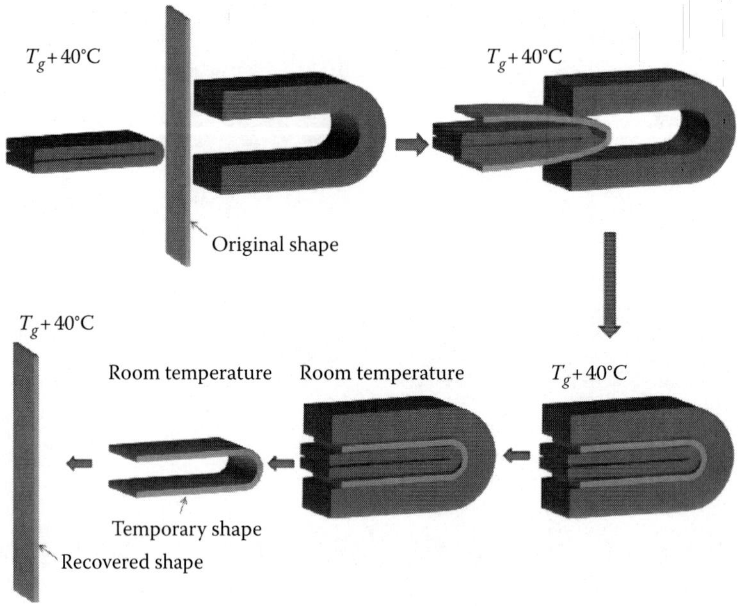

FIGURE 5.19
Schematic representation of the shape memory bending test. (Reprinted from Xie, F. et al., 2014, Synthesis and characterization of high temperature cyanate-based shape memory polymers with functional polybutadiene/acrylonitrile, *Polymer*, 55(23): 5873–5879, with permission from Elsevier.)

an entire shape memory cycle, and the shape recovery is measured in terms of recovery ratio, comparing the original shape and its recovered shape.

To characterize the mechanical and shape memory properties of SMPs, a cyclical thermomechanical test is employed that measures the bend angles (the residual angle of the *m*th thermomechanical bending cycle) using a rod for transformations across the transition temperature.

The value of the shape recovery ratio from the cyclic thermomechanical test (Figure 5.22) is calculated as follows:

$$R_r = \left(\left(\theta_m - \theta_e\right)/\theta_m\right)\times 100\% \tag{5.7}$$

where:

- R_r is the shape recovery ratio of the *m*th thermomechanical bending cycle
- θ_m is the bend angle for a random thermomechanical cycle
- θ_e is the recovery angle

Here, the shape memory rod is bent and recovered around a mandrel, as shown in Figure 5.20, and the bend/recovery angle are noted to calculate the recovery ratio.

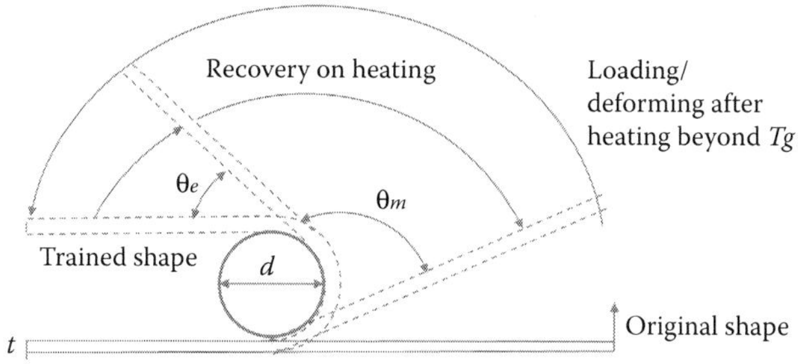

FIGURE 5.20
Cyclical thermomechanical test on an SMP rod of thickness *t*.

Shape memory properties such as the shape recovery ratio (R_r) and shape-fixing ratio (R_f) are explained in detail in Section 5.2. The cyclic behavior of polyimide SMPs has been plotted for thermoresponsive SMPs in Figure 5.21.

Figure 5.21 shows the fatigue of the SMP for *n* cycles, which will establish a relationship between the shape fixity/recovery ratio and efficiency. Consistent recovery at 100% is a good indication of shape memory efficiency. The *shape recovery time* (*t*) is associated with how fast the SMP responds to the stimulus and reaches its recoverable strain in each cycle.

Investigations into shape memory polymer composites (SMPCs) reveal that shape recovery properties strongly depend on the number of

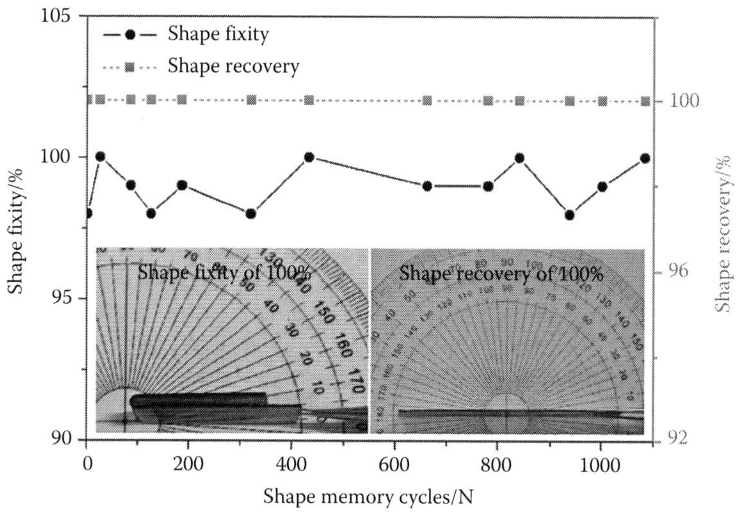

FIGURE 5.21
Shape recovery/fixity percentages vs. the number of bending cycles at $T_g \pm 20°C$ for polyimide-based SMP reinforced with fiber. (Xiao, X. and Kong, D., 2016, High cycle-life shape memory polymer at high temperature, *Scientific Reports*, 33610: 1–10.)

thermomechanical cycles and the recovery properties stabilize with less fluctuations after several cycles (packaging/deployment).

Uniaxial tensile tests determine the properties of SMPs at temperatures above and below T_g. The ultimate strain of the material and the stress associated with straining a material are valuable in designing SMPs. Recovery results show that the SMP can fully recover unrestrained and can exert a force equal to that at which it was strained.

The constant–stress recovery test gives the most informative results, and it can be either dynamic or isothermal when evaluating cyclic behavior.

For dynamic testing, the specimen is heated to above the transition temperature upon unloading the material in the glassy phase. During the entire heating, the material is held at constant stress, and the strain is allowed to begin recovery when it reaches the transition temperature. Upon heating to the desired temperature (T_g+30°C), the temperature is held constant and it completes the recovery.

For isothermal recovery, the material is held at constant strain during the heating process. Due to the strain constraint as it passes the transition temperature, the stress in the material increases. Upon reaching the maximum temperature, the stress is then unloaded to the desired level and held constant. At this point, the strain constraint is released, and the specimen is allowed to recover.

It is possible to obtain the strain recovery profile under a constant stress as well as the maximum recoverable strain. These results will not only help develop the model but will also help determine the limits of the material for designing particular applications.

In contrast to constant–stress recovery experiments, another series of tests include releasing the strained material at a constant strain rate and observing the resulting stress and stress rate trends. Similar to constant–stress recovery tests, constant strain rate experiments are broken into dynamic and isothermal methods.

Dynamic testing is done by heating the material to T_g+30°C after unloading. Until the material reaches the transition temperature, the strain in the material is held constant and the thermal stress increases in the material. As the material passes T_g, it is then unloaded at a constant strain rate. At this point, the stress theoretically relaxes to a minimum value.

In isothermal testing, the material is held at a constant strain until the temperature reaches the maximum temperature. Due to the strain constraint, the material will experience a greater stress increase during its attempt to recover, having passed T_g. Subsequent to reaching the maximum temperature, the material is released at a constant strain rate and stress relaxation occurs.

The major characterization techniques used for morphological studies include Fourier transform infrared (FTIR) spectroscopy, Raman spectroscopy and X-ray diffraction (XRD), optical microscopy, scanning electron microscopy (SEM), and transmission electron microscopy (TEM). Electrical conductivity measurements and temperature distribution measurements are

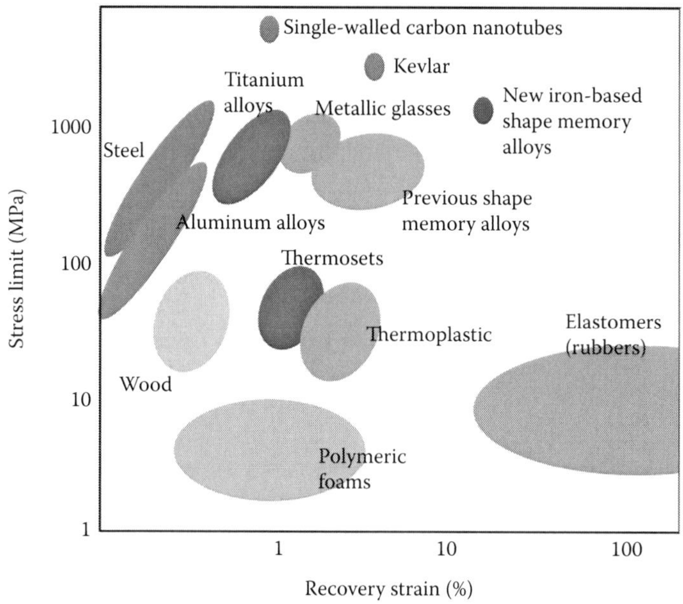

FIGURE 5.22
Stress limit and strain recovery plot to identify optimal materials for SME. (Reprinted from Mohd Jani, J. et al., 2014, A review of shape memory alloy research, applications and opportunities, *Materials and Design*, 56: 1078–1113, with permission from Elsevier.)

detected by infrared camera, and the thermal properties and T_g are obtained by thermogravimetric analysis, DSC, and dynamic mechanical analysis. The mechanical properties, as previously explained, are obtained by tensile and compressive testing. Based on mechanical behavior, smart materials can be plotted on a plane of stress limit and recovery strain, which may help designers and material scientists to choose a suitable material (Figure 5.22).

5.13 Constitutive Models of SMPs

When heated above T_g, the SMP is in a soft rubbery state, with the values of Young's modulus much lower than that of the hard state. A dynamic mechanical thermal analyzer (DMTA) is used to plot the Young's modulus versus the temperature of the SMP. The SMP's response to uniaxial loading represents another common approach to characterization. The higher the cross-linked density, the higher the modulus. A high T_g along with a high-storage modulus results in high stiffness, which in most cases equates to a low percentage of elongation and poor energy dissipation when stressed at ambient temperatures. The modulus of a polymeric shape memory material and the dynamic mechanical properties of a polymer are described in terms of a complex dynamic modulus, defined as

$$E^* = E' + i\,E'' \tag{5.8}$$

where:

- E' is the storage modulus (a measurement of the recoverable strain energy)
- E'' is the loss modulus (the hysteresial energy dissipation)

The viscoelastic properties of SMPs can be obtained from DMA, where a sinusoidal force (stress σ, σ_0) is applied to a material and the resulting displacement (strain ε, ε_0) is measured. For a perfectly elastic solid, the resulting strain and the stress will be perfectly in phase, and for a purely viscous fluid, there will be a 90° phase lag between stress and strain. SMPs modeled as viscoelastic polymers will have the characteristics of solid and viscous materials, and some phase lag will occur.

Thus, when the stress is applied and the strain lags behind,

$$\text{Stress}, \sigma = \sigma_0\, Sin\left(\omega t + \delta\right) \tag{5.9}$$

$$\text{Strain}, \varepsilon = \varepsilon_0\, Sin\left(\omega t\right) \tag{5.10}$$

where:

- ω is the frequency of the strain oscillation
- t is the time
- δ is the phase lag between stress and strain

The storage modulus measures the stored energy, representing the elastic portion, and the loss modulus measures the energy dissipated as heat, representing the viscous portion. The tensile storage and loss moduli are expressed as follows:

Storage modulus:

$$E' = \left(\sigma_0/\varepsilon_0\right) \times Cos\ \delta \tag{5.11}$$

Loss modulus:

$$E'' = \left(\sigma_0/\varepsilon_0\right) \times Sin\ \delta \tag{5.12}$$

Phase angle:

$$\delta = tan^{-1}\left(E''/E'\right) \tag{5.13}$$

It is reported that when deformation is small, the storage modulus is approximately equal to the Young's modulus.

Thermoset SMPs are preferred to thermoplastic SMPs in many cases due to their chemical resistance, higher storage modulus, and better shape memory performance. The shape memory performances of thermoset polyimide SMPs are higher than 98% with R_r and nearly 99% with R_f.

Models based on the finite element method are required to explain shape memory behavior and are significant for engineering analysis and the design of SMP components.

SMPs are modeled macroscopically by many researchers based on empirical relationships, mathematical parameters deduced from the characteristics of materials, or thermodynamic laws. Since shape memory is an entropy-driven mechanism, the use of thermodynamic laws for modeling is realistic, taking into account the mechanical properties. Thus, a blend of mechanical and thermodynamic modeling can predict the shape-changing behavior of SMPs. Note that SMP properties vary depending on many parameters, such as component properties, stimuli, synthesis, and so on. Hence, a general model cannot explain shape memory behavior, which points toward individual model development for each SMP. The rate-dependent and thermomechanical properties of polymers on the macroscopic level are mostly explained with viscoelastic models. In these models, materials are assumed to be the combinations of springs, dashpots, or frictional elements (derived from suggestions by Maxwell, Voigt, or the standard linear viscoelastic models). Accurate two- and three-dimensional models are required to describe and predict the shape memory behavior of pure SMPs, particle-filled SMP composites, and fiber-reinforced SMP composites with stress, strain, time, external stimulus parameters, and filler content.

5.13.1 Maxwell Model

Maxwell's model assumes that viscoelastic material is a combination of a purely viscous damper and a purely elastic spring (Figure 5.23a). The following equation represents the model.

$$\left(d\varepsilon/dt\right)=\left(d\varepsilon_D/dt\right)+\left(d\varepsilon_s/dt\right)=\left(\sigma/\eta\right)+\left(\left(1/E\right)\times\left(d\sigma/dt\right)\right) \tag{5.14}$$

where:

ε	is the net strain
d and s	are the strain in the viscous dashpot/damper and elastic spring, respectively
σ	is the stress applied in the system
η	is the viscosity of the damping portion
E	is the elastic modulus
t	is the time

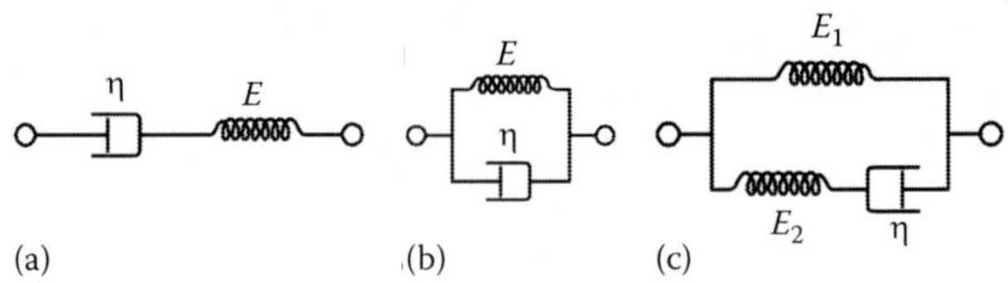

FIGURE 5.23
Viscoelastic spring and dash-pot models. (a) Series connection: Maxwell's model. (b) Parallel connection: Voigt's model. (c) Combination of series and parallel: standard linear solid model.

The equation is derived with consideration of the constant stress on the system and variable strain for each individual element.

The Maxwell model predicts that stress decays exponentially with time, which is accurate for most polymers. One limitation of Maxwell's model is that it does not accurately explain creep.

5.13.2 Kelvin–Voigt Model

The Kelvin–Voigt model (also known as the *Voigt model*) considers the viscoelastic material a parallel combination of a viscous damper and an elastic spring (Figure 5.23b). An advantage of this model over Maxwell's explanation is that this can accurately predict the creep of polymers. But the Voigt model fails to explain the stress relaxation phenomenon in polymers. The following constitutive equation represents the Voigt model of polymeric systems.

$$\sigma(t) = E\varepsilon(t) + \eta\left(d\varepsilon(t)/dt\right) \tag{5.15}$$

The equation is derived with consideration of the constant strain across the system while the applied stress is distributed for each element.

5.13.3 Standard Linear Solid Model

This model effectively combines both Maxwell's and Voigt's models. This has two arms: *Maxwell's arm* and *Voigt's system* (Figure 5.23c). The constitutive equation corresponding to this model is

$$d\varepsilon(t)/dt = (E_2/\eta)\times\left((\eta/E_2)\times\left(d\sigma(t)/dt\right)\right) + E_1\varepsilon(t)\big/(E_1 + E_2) \tag{5.16}$$

where E_1 and E_2 represent the elastic moduli corresponding to both Maxwell's and Voigt's models, the other parameters being the same as in the previous cases.

Even though the standard linear solid model is more accurate than Maxwell's and Voigt's models in respect to material responses, it is difficult to calculate strain under specific loading conditions, which is a drawback of this system of modeling. Modifying these general models based on linearity or thermal behavior can produce specific SMP models that match the results of experimental validations. The continuum theories of rubber elasticity (the Mooney–Rivlin model, proposed by Melvin Mooney in 1940 and expressed in terms of invariants by Ronald Rivlin in 1948) are obeyed by the modeling concepts, which lowers the potential energy for the reaction/process to happen. The Mooney–Rivlin model treats material as a continuous mass rather than as discrete particles.

Among the models based on phase transition, the constitutive model of the uniaxial loading conditions of SMPs uses internal state variables based on experimental results and the molecular mechanisms of SMPs. This model concerns two kinds of extreme phases: the frozen and active phases of SMPs at an arbitrary temperature (Figure 5.24).

The volume fractions of the active and frozen phases are given by

$$\Phi_f = \left(V_{\text{frz}}/V\right) \tag{5.17}$$

$$\Phi_a = \left(V_{\text{act}}/V\right) \tag{5.18}$$

$$\Phi_f + \Phi_a = 1 \tag{5.19}$$

where:

Φ_f and Φ_a represent the fractions of the frozen and active phases in the SMP, respectively

V, V_{frz}, and V_{act} are the total volume and the volumes of the frozen phase and the active phase, respectively

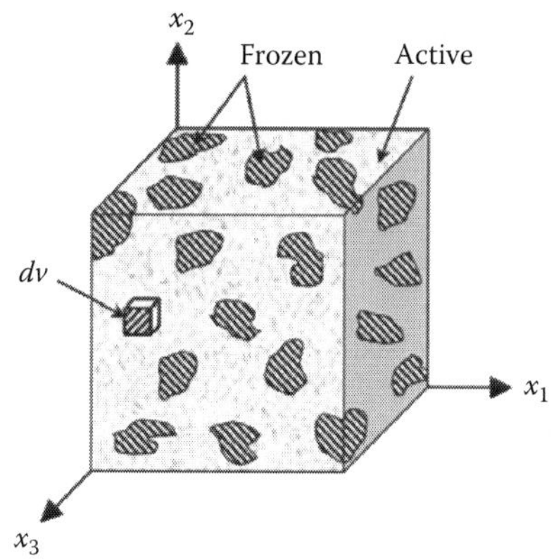

FIGURE 5.24
Three-dimensional SMP constitutive model based on uniaxial loading conditions showing two extreme phases representing a polymer in the glass transition state with a predominant active phase. (Reprinted from Hu, J. et al., 2012, Recent advances in shape memory polymers: Structure, mechanism, functionality, modeling and applications, *Progress in Polymer Science*, 37(12): 1720–1763, with permission from Elsevier.)

The three components of the strain in the frozen phase are the stored strain (ε_s), mechanical elastic strain (ε_m), and thermal expansion strain (ε_T) and are expressed as

$$\varepsilon = \varepsilon_s + \varepsilon_m + \varepsilon_T \tag{5.20}$$

The stored strain of an SMP is given by

$$\varepsilon_s = \text{Int } \left(0 \text{ to } \Phi_f\right)\left(\varepsilon_{ef}\left(X\right)d\Phi\right) \tag{5.21}$$

where:

- ε_{ef} is the entropic frozen strain
- X is the position vector of the frozen phase in the anisotropic segment
- Int is the integration of the function with limits inside the parenthesis

Note that shape memory in SMAs is enthalpy driven, and in SMPs it is entropy driven. During deformation, the chains are aligned/stretched (as explained in Section 5.7) and the configurational entropy decreases.

The mechanical elastic strain (based on Hook's law of elasticity) of an SMP is given by the following expression:

$$\varepsilon_m = \left(\Phi_f S_i + \left(1 - \Phi_f\right) S_e\right) \times \sigma \tag{5.22}$$

where:

- σ is the second-order stress tensor
- S_i is the elastic compliance corresponding to the internal energetic deformation
- S_e is the elastic compliance corresponding to the entropic deformation

The thermal expansion strain is expressed as

$$\varepsilon_T = \text{Int } \left(T_0 \text{ to } T\right)\left(\Phi_f \alpha_f + \left(1 - \Phi_f\right)\alpha_a\right)dT_i \tag{5.23}$$

where:

- α_f and α_f are the thermal expansion coefficients of the frozen phase and the active phase
- T_i is the instantaneous temperature

Thus, Equation 5.20 can be rewritten as

$$\varepsilon = \text{Int}\left(0 \text{ to } \Phi_f\right)\left(\varepsilon_{ef}\left(X\right)d\Phi + \left(\Phi_f S_i + \left(1 - \Phi_f\right)S_e\right)\sigma\right) + \text{Int}\left(T_0 \text{ to } T\right)$$
$$\left(\Phi_f \alpha_f + \left(1 - \Phi_f\right)\alpha_a\right)dT_i \tag{5.24}$$

Since Young's modulus $E = (\sigma/\varepsilon_m)$ (Hook's law), the constitutive stress value (σ) for the overall volume of the polymer is

$$\sigma = E\text{Int}\left(0 \text{ to } \Phi_f\right)\left(\left(\varepsilon_{ef}(x)d\Phi + \left(\Phi_f S_i + \left(1-\Phi_f\right) S_e\right)\sigma\right) + \text{Int}\left(T_o \text{ to } T\right)\right.$$
$$\left.\left(\Phi_f\alpha_f + \left(1-\Phi_f\right)\alpha_a\right)dT_i\right) \tag{5.25}$$

Thus, stress is expressed as a function of strain, temperature, the frozen fraction, and the frozen entropic strain. By simplifying Equation 5.25, it becomes

$$\sigma\left(\varepsilon, T, \varepsilon_{ef}, \Phi_f\right) = E\left(\varepsilon_m + \varepsilon_s + \varepsilon_T\right) \tag{5.26}$$

This theoretical modeling of mechanical constitutive behavior shows that it is difficult to model an SMP closely to reality due to the large recoverable deformation and the heavy dependence of polymer properties on time and temperature.

The deployment dynamics of SMP composite space structures have been studied mathematically, and their formulations combine the constitutive laws of SMP composites, nonlinear strain–displacement relations, and the equations of motion. This concept is also anchored to the presence of a frozen phase and an active phase in an SMP. There is an unfortunate lack of SMP models, in spite of SMPs being a good replacement for SMAs and many other smart materials. Several macroscopic or phenomenological constitutive models and microscopic or physical models explain SMP behavior in small and finite deformation regimes.

By having a thorough understanding of the basic concepts of material properties, smart structures can be tailored for specific applications. For structural materials, the stress limit and recoverable strain form the deciding factors, the comparison of which supports material selection for smart shape recovery structures. Referring to Figure 5.22, elastomeric materials demonstrate the maximum recoverable strain, while single-walled CNTs have the maximum stress limits. The basics of composite materials show that a proper blend of the relevant materials can enhance the capabilities of the resultant product. The idea of mixing carbon nanomaterials (CNTs, CB, etc.) to enhance their mechanical properties in addition to their electrical and thermal properties comes from this insight and in turn draws attention toward composite materials with shape memory properties.

References

Deng, Z., Guo, Y., Zhao, X., Li, L., Dong, R., Guo, B., and Ma, P. X., 2016. Stretchable degradable and electroactive shape memory copolymers with tunable recovery temperature enhance myogenic differentiation, *Acta Biomaterilia*, 46: 234–244.

Hager, M. D., Bode, S., Weber, C., and Schubert, U. S., 2015. Shape memory polymers: Past, present and future developments, *Progress in Polymer Science*, 49: 3–33.

Hu, J., Zhu, Y., Huang, H., and Lu, J., 2012. Recent advances in shape-memory polymers: Structure mechanism, functionality, modeling and applications, *Progress in Polymer Science*, 37(12): 1720–1763.

Huang, W. M., Zhao, Y., Wang, C. C., Ding, Z., Purnawali, H., Tang, C., and Zhang, J. L., 2012. Thermo/chemo-responsive shape memory effect in polymers: A sketch of working mechanisms, fundamentals and optimization, *Journal of Polymer Research*, 19(9): 952–986.

Leng, J., Lan, X., Liu, Y., and Du, S., 2011. Shape-memory polymers and their composites: Stimulus methods and applications, *Progress in Materials Science*, 56(7): 1077–1135.

Liu, C., Qin H., and Mather, P. T., 2007. Review of progress in shape-memory polymers, *Journal of Materials Chemistry*, 17(16): 1543–1558.

Luo, X. and Mather, P. T., 2010. Conductive shape memory nanocomposites for high speed electrical actuation, *Soft Matter*, 6(10): 2146.

Luo, X. and Mather, P. T., 2013. Shape memory assisted self-healing coating, *ACS Macro Letters*, 2: 152–156.

Mohd Jani, J., Leary, M., Subic, A., and Gibson, M. A., 2014. A review of shape memory alloy research, applications and opportunities, *Materials and Design*, 56: 1078–1113.

Sinn-Hanlon, J., White, S., and Blaiszik, B., 2012. Release of healing agents. Retrieved from https://sites.google.com/site/shmolfw/design-strategies/release-of-healing-agents.

Sun, L., Huang, W. M., Ding, Z., Zhao, Y., Wang, C. C., Purnawali, H., and Tang, C., 2012. Stimulus-responsive shape memory materials: A review, *Materials and Design*, 33(1): 577–640.

Vernon, L. B. and Vernon, H. M., 1941. Process of Manufacturing Articles of Thermoplastic Synthetic Resins, United States Patent 2,234,993, Cl 18–55.

Xiao, X. and Kong, D., 2016. High cycle-life shape memory polymer at high temperature, *Scientific Reports*, 33610: 1–10.

Xie, F., Huang, L., Liu, Y., and Leng, J., 2014. Synthesis and characterization of high temperature cyanate-based shape memory polymers with functional polybutadiene/acrylonitrile, *Polymer*, 55(23): 5873–5879.

Zhang, H., Wang, H., Zhong, W., and Du, Q., 2009. A novel type of shape memory polymer blend and the shape memory mechanism, *Polymer*, 50(6): 1596–1601.

Zhang, F., Zhang, Z., Ion, T., Liu, Y., Leng, J., 2015. Shape memory polymer nanofibers and their composites: electrospinning, structure, performance and application, *Frontiers in Materials*, 2(Oct): 1–10.

Bibliography

Behl, M. and Lendlein, A., 2007. Shape-memory polymers, *Materials Today*, 10(4): 20–28.

Biju, R. and Reghunadhan Nair, C. P., 2013. Synthesis and characterization of shape memory epoxy-anhydride system, *Journal of Polymer Research*, 20(2): 82p.

Ding, Z., 2012. Shape memory hybrids: Mechanism and design for tailored properties, postgraduate thesis work, School of Mechanical and Aerospace Engineering, Nanyang Technological University, Singapore.

Dyana Merline, J. and Reghunadhan Nair, C. P., 2012. Carbon/epoxy resin based elastic memory composites, *Eurasian Chemico-Technological Journal*, 14(3): 227–232.

Flory, P. J. and Rehner, J., Jr., 1943. Statistical mechanics of crosslinked polymer networks, *Journal of Chemical Physics*, 11, 512p.

García-Huete, N., Axpe, E., Cuevas, J. M., Mérida, D., Laza, J. M., García, J. A., Vilas, J. L., Plazaola, F., León, L. M., 2017. In situ measurements of free volume during recovery process of a shape memory polymer, *Polymer*, 109: 66–70.

Gross, K. E., 2008. Mechanical characterization of shape memory polymers to assess candidacy as morphing aircraft skin, postgraduate thesis work, University of Pittsburgh, PA.

Han, Z. and Fina, A., 2011. Thermal conductivity of carbon nanotubes and their polymer nanocomposites: A review, *Progress in Polymer Science (Oxford)*, 36(7): 914–944.

Hartl, D. J. and Lagoudas, D. C., 2007. Aerospace applications of shape memory alloys, *Journal of Aerospace Engineering*, 221: 535–552.

Huang, W. M., 2013. Shape memory polymers (SMPs): Current research and future applications, *AZO Materials*, 1–7.

Jiang, Q., Wang, X., Zhu, Y., Hui, D., and Qiu, Y., 2014. Mechanical, electrical and thermal properties of aligned carbon nanotube/polyimide composites, *Composites Part B: Engineering*, 56: 408–412.

Kuksenok, O., Balazs, A. C., Liu, Y., Harris, V., Nan, H. Q., Mujica, M., and Vasquez, Y., et al., 2016. Stimuli-responsive behavior of composites integrating thermo-responsive gels with photo-responsive fibers, *Materials Horizons*, 3(1): 53–62.

Lan, X., Leng, J. S., Liu, Y. J., and Du, S. Y., 2008. Investigate of electrical conductivity of shape-memory polymer filled with carbon black, *Advanced Materials Research*, 47–50: 714–717.

Leng, J., Lu, H., and Du, S., 2008. Conductive shape memory polymer composite technology and its applications in aerospace, *Proceedings of 49th AIAA/ASME/ASCE/AHS/ASC Structures, Structural Dynamics, and Materials Conference; 16th AIAA/ASME/AHS Adaptive Structures Conference*, Schaumburg, IL, April, pp. 1–8.

Leng, J., Lu, H., Liu, Y., Huang, W. M., and Du, S., 2009. Shape-memory polymers: A class of novel smart materials, *MRS Bulletin*, 34(11): 848–855.

Leng, J. and Ye, L., 2009. Smart materials and nanocomposites: Bring composites to the future, *Composites Science and Technology*, 69(13): 2033.

Lin, J., 2006. Shape memory rigidizable inflatable (RI) structures for large space systems applications, *Proceedings of 47th AIAA/ASME/ASCE/AHS/ASC Structures, Structural Dynamics, and Materials Conference*, Newport, RI, May, pp. 1–11.

Liu, Y., Lv, H., Lan, X., Leng, J., and Du, S., 2009. Review of electro-active shape-memory polymer composite, *Composites Science and Technology*, 69(13): 2064–2068.

Liu, Y., Wang, X., Lan, X., Lv, H., and Leng, J., 2008. Shape memory polymer composite and its application in deployable hinge for space structure, *Proceedings of SPIE 2008: Sensors and Smart Structures Technologies for Civil, Mechanical, and Aerospace Systems*, 6932(2): 10–18.

Luo, X. and Mather, P. T., 2013a. Design strategies for shape memory polymers., *Current Opinion in Chemical Engineering*, 2(1): 103 111.

Ma, L., Zhao, J., Wang, X., Chen, M., Liang, Y., Wang, Z., Yu, Z., and Hedden, R. C., 2015. Effects of carbon black nanoparticles on two-way reversible shape memory in crosslinked polyethylene, *Polymer (United Kingdom)*, 56: 490–497.

Marconnet, A. M., Yamamoto, N., Panzer, M. A., Wardle, B. L., and Goodson, K. E., 2011. Thermal conduction in aligned carbon nanotube–polymer nanocomposites with high packing density, *American Chemical Society Nano*, 5.6: 4818–4825.

Mather, P. T., Luo, X., and Rousseau, I. A., 2009. Shape memory polymer research, *Annual Review of Materials Research*, 39: 445–471.

Merline, J. D., Reghunadhan Nair, C. P., and Ninan, K. N., 2008. Synthesis, characterization, curing and shape memory properties of epoxy-polyether system, *Journal of Macromolecular Science, Part A: Pure and Applied Chemistry*, 45(4): 312–322.

Michaud, V., 2004. Can shape memory alloy composites be smart? *Scripta Materialia*, 50(2): 249–253.

Quan, D. and Hai, X., 2015. Shape memory alloy in various aviation field, *Procedia Engineering*, 99: 1241–1246.

Ragin Ramdas, M., Santhosh Kumar, K. S., and Reghunadhan Nair, C. P., 2015. Synthesis, structure and tunable shape memory properties of polytriazoles: Dual-trigger temperature and repeatable shape recovery, *Journal of Materials Chemistry A*, 3(21): 11596–11606.

Sokolowski, W. M., 2004. US Patent no. 006702976 B2: Cold hibernated elastic memory self-deployable and rigidizable structure and method therefor, March 9.

Sokolowski, W. M. and Hayashi, S., 2003. Applications of cold hibernated elastic memory (CHEM) structures, *Proceedings of SPIE: 10th International Symposium on Smart Structures and Materials: The International Society for Optical Engineering*: 5056: 534–544.

Sunitha, K., Santhosh Kumar, K. S., Mathew, D., and Reghunadhan Nair, C. P., 2013. Shape memory polymers (SMPs) derived from phenolic cross-linked epoxy resin via click chemistry, *Materials Letters*, 99: 101–104.

Van Humbeeck, J., 1999. Non-medical applications of shape memory alloys, *Materials Science and Engineering: A*, 273–275: 134–148.

Wei, Z. G., Sandstrom, R., and Miyazaki, S., 1998. Shape-memory materials and hybrid composites for smart systems, part I: Shape-memory materials, *Journal of Materials Science*, 33(15): 3743–3762.

Wu, X., Huang, W. M., Zhao, Y., Ding, Z., Tang, C., and Zhang, J., 2013. Mechanisms of the shape memory effect in polymeric materials, *Polymers*, 5(4): 1169–1202.

Xiao, X., Qiu, X., Kong, D., Zhang, W., Liu, Y., and Leng, J., 2016. Optically transparent high temperature shape memory polymers, *Soft Matter*, 12(11): 2894–2900.

Xie, F., Huang, L., Leng, J., and Liu, Y., 2016. Thermoset shape memory polymers and their composites, *Journal of Intelligent Material Systems and Structures*, 27(18): 2433–2455.

Xie, T., 2011. Recent advances in polymer shape memory, *Polymer*, 52(22): 4985–5000.

Yang, D., 2000. Shape memory alloy and smart hybrid composites: Advanced materials for the 21st century, *Materials and Design*, 21(6): 503–505.

6

Shape Memory Hybrids

As discussed in previous chapters, shape memory materials have the ability to regain their original shapes from a temporary/trained shape when a particular stimulus is applied. Shape memory alloys (SMAs), ceramics (SMCs), gels (SMGs), and polymers (SMPs) are individual materials that exhibit the property of shape memory, whereas an intermediate property requirement can be met by a suitably designed combination of any of these materials. Consider the case of reinforced cement concrete, a composite civil engineering construction material. It is a combination of Portland cement (a binder material), stone aggregate (the bulk component), steel reinforcement (for taking tensile loads), and water (to initiate the hydration reaction and strengthen it). Thus, individual material properties are combined in design ratios to serve a specific purpose. Similarly, a combination of several materials with entirely different properties forms a hybrid material that can suit specific applications in precise actuation, combining the advantages of both the materials.

Details of smart shape-memorizing hybrids are explained in brief in the following sections.

6.1 Hybrid Materials with Shape Memory Properties

As the name suggests, hybrid composite materials consist of one or more smart ingredients as either matrix or filler. Smart systems are not single-component materials; rather, they represent integrated systems of materials. Thus, the mechanism of shape memory is analogous to that in SMPs, which are explained in detail in Chapter 5. Shape memory hybrids (SMHs) are also dual-domain systems in which one phase is the elastic domain while the other is the transition domain. One of the major differences between SMP composites and SMHs is that, component-wise, SMHs may not behave as shape memory materials, but the presence/combination of the elastic and transition domains causes the shape memory effect (SME) upon reaching the transition temperature. Thin films, fibers, wires, particle fillers, and bulk matrices are a few forms of shape memory materials that can be incorporated with other materials with desirable properties to form hybrid shape memory materials.

6.2 Mechanism and Concept

We saw in Chapter 2 that SME in SMAs is due to the reversible martensitic transformation as the driving mechanism. Chapter 4 explains that SME in polymers results from a two-segment system—a switching domain known as the *soft phase* and a fixed domain known as the *hard phase*. Similarly, SMHs possess two components known as the *elastic segment* (analogous to the switching domains of SMPs) and the *transition segment* (the fixed/frozen/hard domain of SMPs). Upon application of suitable stimuli, the elastic segment causes shape transformation, whereas the transition segment "remembers" the original shape to recover.

Material selection for SMAs, SMCs, SMGs, and SMPs is as critical as the combining ratio (the ratio of quantities of nickel to titanium in NiTinol decides the shape memory property of the NiTinol SMA) or the number of soft and hard phases (in the case of SMPs). But the materials selected for both components of SMHs need not possess SME as an individual property. The thermomechanical properties of SMHs are based on the amount and behavior of individual domains. The type of stimulus can be decided from the transition domain chosen, whereas the elastic domain can provide the required stiffness and amount of shape recovery in an SMH.

A mechanism with an SMH made of polymeric material without SME behavior combined with an SMA spring in a system is illustrated in Figure 6.1.

Figure 6.1 shows the self-healing mechanism of an SMH polymeric system with an SMA spring embedded inside a sheath made of a hybrid material. Upon stretching the element by applying an external load, the outer sheath breaks apart (Figure 6.1b1–b2), while the inner SMA wire extends/deforms. This system is heated to the transition temperature (Figure 6.1b3–b4) to retrieve the original configuration of the SMA spring. The broken segments

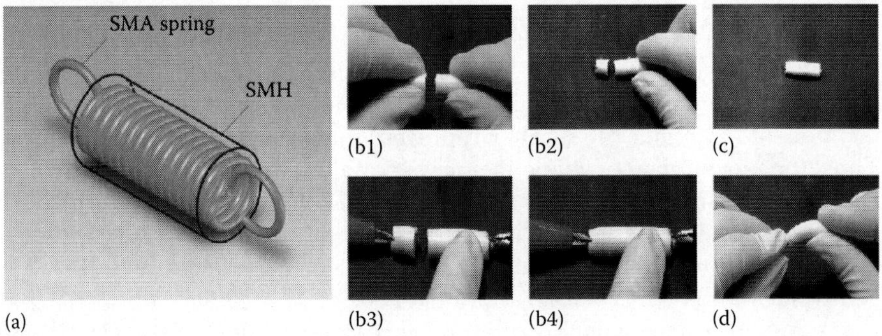

FIGURE 6.1
Depiction of the self-healing of SMH. (a) SMA spring inside SMH sheath. (b1–b2) Fracture made to demonstrate self-healing behavior. (b3–b4) Shape recovery stimuli as an electric current for Joule heating. (c–d) Self-healing of the sample and checking the integrity of the healed sample. (Reprinted from Huang, W. M. et al., 2010, Shape memory materials, *Materials Today*, 13(7–8): 54–61, with permission from Elsevier.)

on the SMH's outer sheath touch and join together again to heal the gap formed (the shape memory property of the hybrid leads to self-healing) (Figure 6.1c–d). Thus, the hybrid SMA and polymer/gel material goes back to its integral form once it has healed. SMHs do not require a curing agent for self-healing, unlike individual SMPs, and remain stable for numerous cycles.

It has been reported that chemical interactions between components, both elastic and transition, should be kept to a minimum or completely avoided. This is to facilitate the precise prediction of the properties and behaviors of the SMHs from the known individual components. There are many ways this is possible, one of which is to select inert components and another is to use a physical barrier.

One published work (Xuelian et al. 2013) has reported an SMH made of a polymeric sponge and a water temperature–sensitive gel, P407 (a material approved by the US Food and Drug Administration for biomedical applications), whose structure consists of a hydrophobic block of poly(propylene oxide) and two hydrophilic blocks of poly(ethylene oxide) (PEO–PPO–PEO). P407 aqueous solution has a thermally reversible property by which it melts when the temperature is lowered from 35°C to 4°C.

Similar to the heat-responsive shape memory materials discussed previously, cooling is complementary to the SMH (as demonstrated in Figure 6.2), which makes use of the elastic component (a polymeric sponge in a tube shape serving as matrix material) and P407's thermal properties. In this SMH, P407 is considered the transition component and human body temperature the triggering stimulus. Figure 6.3 depicts the water responsiveness of SMHs that make use of the hydrophilic and hydrophobic components of P407 to recover their shape in 0°C water and 20°C water.

Upon immersion in 0°C water, the shape is recovered in less than 1 min, which can be explained by the melting of the P407 gel upon cooling (Figure 6.3).

As with SMAs, SMPs, and so on, the recovery strain, stress, and speed of recovery are significant in the case of SMHs and are being studied/experimented with by material scientists around the world. Similarly, the rationale behind the development of hybrids is the ease of processing and the

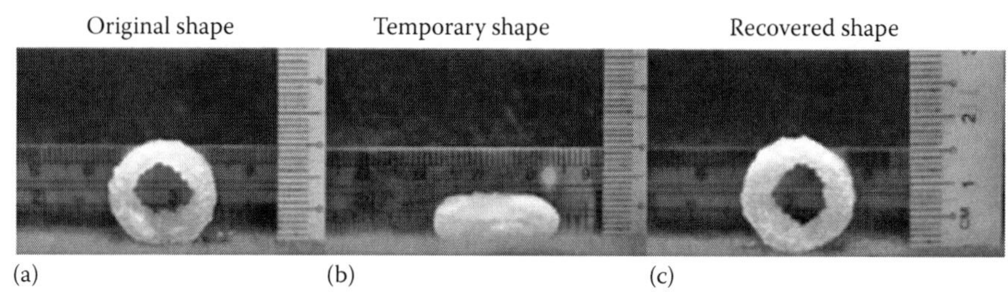

FIGURE 6.2
Demonstration of the P407–sponge SMH cooling response through the recovery of the original shape. (Reprinted from Wang, C. C. et al., 2012, Cooling-/water-responsive shape memory hybrids, *Composites Science and Technology*, 72(10): 1178–1182, with permission from Elsevier.)

FIGURE 6.3
Shape recoveries of the water-responsive P407–sponge SMH at 0°C at 10, 25, 50, and 1140 s. The 90% recovery observed is achieved within 1 min. (Reprinted from Wang, C. C. et al., 2012, Cooling-/water-responsive shape memory hybrids, *Composites Science and Technology*, 72(10): 1178–1182, with permission from Elsevier.)

unlimited possibilities of material selection (as no specific shape memory property is a requisite). Mechanical properties such as conductivity and temperature sensitivity are other factors that bring SMHs to the research table.

References

Huang, W. M., Ding, Z., Wang, C. C., Wei, J., Zhao, Y., and Purnawali, H., 2010. Shape memory materials, *Materials Today*, 13(7–8): 54–61.

Wang, C. C., Huang, W. M., Ding, Z., Zhao, Y., and Purnawali, H., 2012. Cooling-/water-responsive shape memory hybrids, *Composites Science and Technology*, 72: 1178–1182.

Xuelian, W., Huang, W. M., Zhao, Y., Ding, Z., Tang, C., and Zhang J., 2013. Mechanisms of the shape memory effect in polymeric materials, *Polymers*, 5: 1169–1202.

Bibliography

Fan, K., Huang, W. M., Wang, C. C., Ding, Z., Zhao, Y., Purnawali, H., Liew, K. C., and Zheng, L. X., 2011. Water-responsive shape memory hybrid: Design concept and demonstration, *Express Polymer Letters*, 5(5): 409–416.

Huang, W. M., Yang, B., Qing, F. Y., 2017. *Polyurethane Shape Memory Polymers*, CRC Press, Boca Raton, FL.

Van Humbeeck, J., 1999. Non-medical applications of shape memory alloys, *Materials Science and Engineering: A*, 273–275: 134–148.

7

Shape Memory Polymer Composites

Composite materials are those in which two or more different materials have been macroscopically combined to create a superior and unique material with characteristics that differ from the individual components.

7.1 Composite Materials

Composite materials have been popular since the age of Mongolian bows, but they have been in use since time immemorial. Composites are broadly classified into two categories: *particle reinforced* and *fiber reinforced*. Structural composites are entirely different from the other two types, as they are formed by the laminar arrangement of different materials with different properties to imbue the qualities of each in the final product. One example of this is a sandwich structure with an aluminum honeycomb core used for lightweight panels to replace heavy steel panels of equal strength, where the core gives the required stiffness and the outer skins give the finish and protection to the core. Fiber-reinforced composite materials (Figure 7.1a–e) are popular among these types due to the obvious advantages of their intrinsic strength along with their predictable load transfer mechanisms. Fiber-reinforced composite materials consist of two parts: the *matrix* and the *reinforcements*. The matrix is the surrounding medium supporting the reinforcements that protects and transfers the load in the desired directions.

Some of the most common classifications of fiber-reinforced composite materials based on matrix constituents are polymer/organic matrix composites (PMCs/OMCs), metal matrix composites (MMCs), and ceramic matrix composites (CMCs). The obvious advantages of switching to composite materials from a conventional single material product are their high strength-to-weight ratios, resistance to corrosion, controllable conductivity and flexibility, and high durability and reliability.

Because polymer matrices are easier to fabricate, easier to characterize, and more readily available, more recently, the term *composite* has come to describe reinforced plastics, which claim the lion's share of materials used in aerospace applications. Polyester, vinyl ester, epoxy, phenolic, polyimide, polyamide, polypropylene, and polyether ether ketone (PEEK) are a few matrix materials in use, while common fibers used for reinforcements include glass fibers, carbon fibers, aramid fibers, silicon carbide fibers, and cellulose.

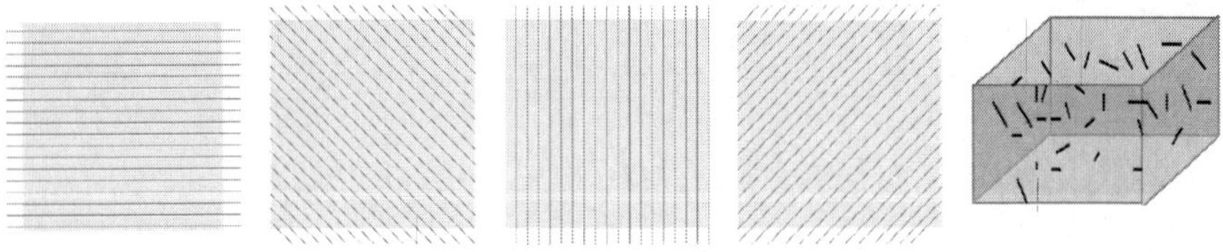

FIGURE 7.1
Fiber-reinforced polymer composites. (a), (b), (c), (d) Different orientations (90°, 45°, 0°, −45°) of long fibers in matrix. (e) Randomly oriented short fibers in matrix.

7.2 Shape Memory Polymer Composites

Looking at aerospace applications, most of the launch vehicles and satellites in use today are made of composite materials due to their aforementioned advantages. For obvious reasons (research for alternatives is in progress), few parts are still made of steel structures and few mechanisms are very heavy to begin with. Mechanisms or actuators made of mechanically movable parts contribute to the complexity and weight of such structures. As previously discussed, the actuations of shape memory polymers (SMPs) can potentially replace mechanisms. Combining one or more materials to SMPs and retaining the shape memory behavior will create shape memory polymer composites (SMPC). Having a complete understanding of the behavior of SMPCs as alternative materials for existing "mechanisms" will benefit the search for improvements in rocketry and spacecraft.

Mechanical hinges, stored energy devices (pyrosystems), and motor-driven tools are currently the most popular and reliable mechanisms used for configuration changes and the deployment and detachment of structures in rockets and in-orbit spacecraft. The intrinsic drawbacks of such traditional devices are their complex assembly processes, multifarious mechanisms, large volumes/weights, and undesired effects such as shocks during deployment.

An SMM as a composite material that forms a part to replace such mechanisms can overcome this bottleneck. Thus, SMPCs are significant enough to transform conventional mechanisms to smart materials systems—for example, a single planar material trained to remember another configuration, replacing all gear spring relay systems currently in use. The general understanding that fewer movable parts gives rise to greater system reliability is significant throughout this.

Pure SMPs exhibit low strength and stiffness in general and an elastic modulus at and around 10 MPa. To improve these properties for diverse applications, the incorporation of reinforcement fillers has been investigated by scientists across the globe. The effects of adding reinforcements to SMPs to make them SMPCs are discussed in this section.

In comparison with metals and ceramics, which are competent matrix materials for composites, reinforced SMP composites have higher specific strength, resistance to corrosion, and toughness. But the low stiffness of SMP resins results in a relatively small recovery force under constraints compared with alternative active actuation materials or schemes. Since the variables that determine the parameters of recovery strain are dependent on stimuli, fiber content, and other characteristics, research in this area could give the edge to SMPCs over other matrix materials in terms of recovery force.

Note that concrete as a construction material is strong in compression is weak in tension. The introduction of steel reinforcements that are strong in tension makes *reinforced concrete* a comprehensively strong material. Analogously, the addition of reinforcements to an SMP matrix helps to overcome its drawbacks by tailoring the material properties. Carbon, fiberglass, and Kevlar reinforcements increase the stiffness of the SMP resins and improve recoverable strain levels.

7.2.1 Modulus, T_g, and Shape Recovery

Discontinuous fiber-reinforced composites exhibit isotropic behavior in shape recovery due to their distribution, while continuous long fiber-reinforced composites are anisotropic (recoverability under transverse tension and bending). The *modulus of elasticity* or *Young's modulus,* a measure of the stiffness of the material and that of a composite, is a factor of the component stiffness. For SMPCs, the stiffness and recoverable strain levels depend strongly on the volume fraction of the discontinuous reinforcement. As per Leng et al. (2011), it is revealed that the addition of a 50% volume fraction of chopped glass fiber increased the composite stiffness by a factor of 4.0 and decreased the recoverable strains by a factor of about 2.5, thus establishing the significance of SMPCs over SMPs for applications.

It is interesting to note the changes brought about in the system by the addition of reinforcements. Comparing the base resin and SiC reinforced resin in Figure 7.2, the storage modulus drops continually from 25°C, while the loss modulus begins to decrease around 60°C. The rate of the decrease of the storage modulus markedly increases as the temperature approaches 60°C. The onset of the *tan delta* (tan δ) curve for both materials appears around 30°C. The peak of either the modulus curve or the peak of the tan δ curve is often used to define the glass transition temperature (T_g), which is reduced by the addition of the reinforcement.

Although the onset temperatures are similar in the reinforced and unreinforced SMP materials, the transition occurs faster in the reinforced material as the tan δ curve narrows. The addition of nanoparticulate SiC reinforcements increases the hardness as well as the elastic modulus of the base resin

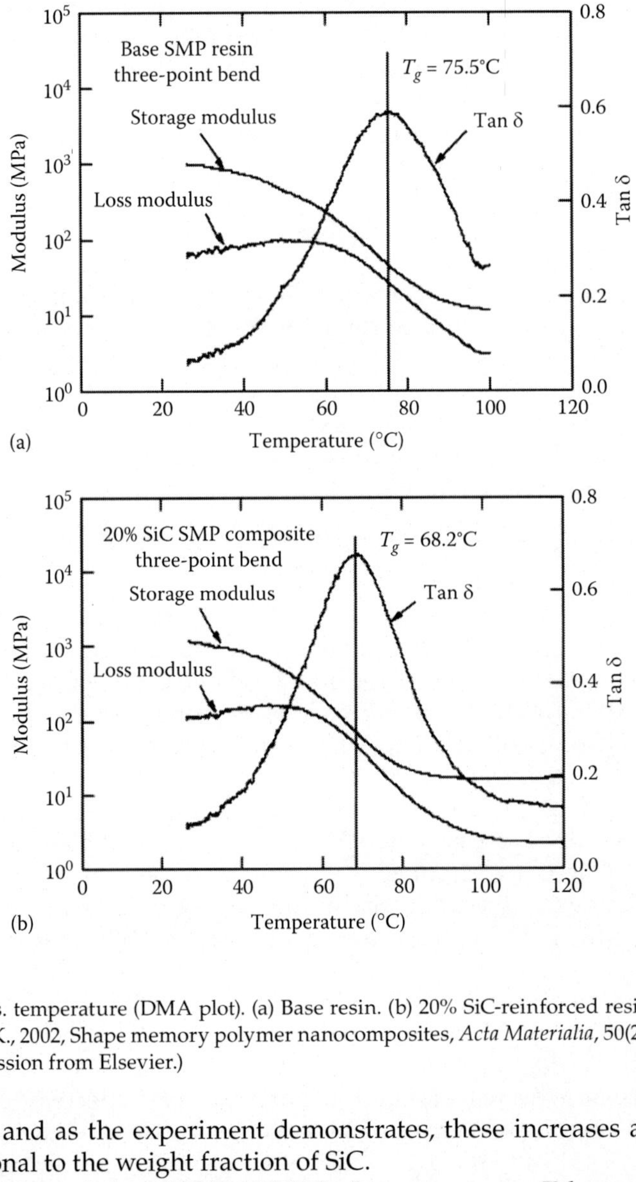

FIGURE 7.2
Modulus vs. temperature (DMA plot). (a) Base resin. (b) 20% SiC-reinforced resin. (Reprinted from Gall, K., 2002, Shape memory polymer nanocomposites, *Acta Materialia*, 50(20): 5115–5126, with permission from Elsevier.)

material, and as the experiment demonstrates, these increases are directly proportional to the weight fraction of SiC.

The elastic modulus of SMPs/SMPCs decreases across T_g by several orders of magnitude, as shown in the research (Gall, 2002). The T_g of a composite is decided by the T_g of the matrix, and little effect on the T_g of the fabric is found in SMPC behavior. Any increase in T_g observed while increasing the reinforcement content is attributed to the resin at the interface of the system. Generally with composites, the interface has a different composition from the bulk matrix, which plays a major role in the failure modes of composites, which also applies to SMPCs. Filler materials have a significant effect on the stiffness, which in turn affects the efficiency of the shape memory behavior, as shown in Figure 7.3.

The shape recovery angle (as discussed in Chapter 5, Section 5.12.4, "Cyclic/Bending Tests") is plotted as a function of time for an SMP and an SMP–carbon black (CB) composite, as shown in Figure 7.3. The addition of CB increases the recovery force and reduces the recovery time. It also enhances the SMP–CB composite's ability to absorb infrared light (this particular SMP is light stimulated). Thus, it is evident that SMPCs perform better than the corresponding SMPs, due to the effects of the reinforcements. Quantification of such enhanced properties is required to suit it to different applications. Similarly, the presence of SiC in various weight fractions and its effects on the recovery speed have been studied (Gall, 2002). Figure 7.4 gives an insight into the amount that reinforcements influence the shape memory property.

Clearly, the recoverability of the composite depends on the type and volume fraction of the reinforcement. The addition of SiC lowers the unconstrained recoverable strain limit and increases the attainable constrained recovery stress. Volume fractions from 0% to 40% give various rates of recovery, and complete recovery is not observed in 40% SiC content after 20 s. This may be an indication that there is an optimal reinforcement content that decides the shape memory behavior. An increase beyond the optimal range may cause fiber–fiber interaction or reduce the role of the matrix in shape recovery.

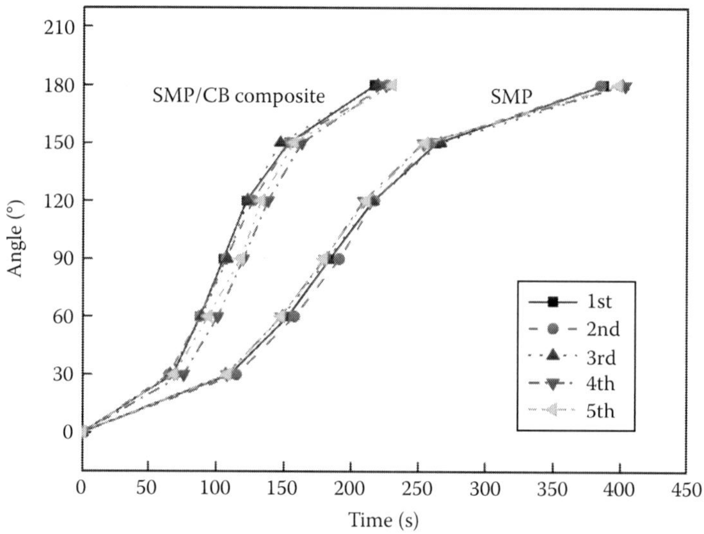

FIGURE 7.3
Recovery angle as a function of time for five tests of an SMP and an SMP–CB composite. (Reprinted from Leng, J. et al., 2011, Shape-memory polymers and their composites: Stimulus methods and applications, *Progress in Material Science*, 56(7): 1077–1135, with permission from Elsevier.)

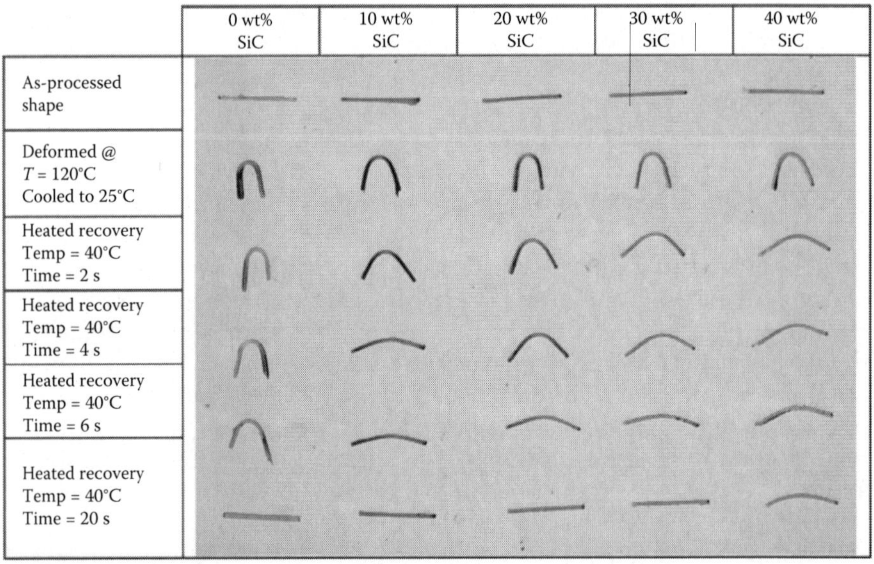

FIGURE 7.4
Unconstrained bend recoverability as a function of SiC fraction and time at temperature. (Reprinted from Gall, K. et al., 2002, Shape memory polymer nanocomposites, *Acta Materialia*, 50(20): 5115–5126, with permission from Elsevier.)

7.3 Nanomaterial-Reinforced Polymer Composites

Similar to organic or inorganic fibers of considerable cross-sectional dimensions, nanofibers/microfibers/particles are used to reinforce polymer matrices with shape memory behavior. This is primarily intended to improve their electrical or thermal conductivity or to introduce their response behavior to specific stimuli such as chemicals. SMPs with reinforced with nanomaterials are referred to as *nanocomposites*. Notably, polymer matrices are highly affected by the dimensions, dispersion states, and interactions of the nano-sized reinforcements play significant roles. An important attribute of SMP nanocomposites, as pointed out in much of the literature (see Leng et al., 2011; Liu et al., 2014; and Mather et al., 2009, to name a few), is the trade-off between recoverable deformation and the external force that causes the deformation.

A recent development is the introduction of a very small amount of nano-sized carbon fillers (3.3 wt % multi-walled carbon nanotubes [MWCNTs]) into polyurethane SMPs to form SMPCs with recovery stress increased up to four times.

Polyurethane SMPs reinforced with combinations of CNTs and CB show increased shape fixity with almost 100% shape recovery, whereas polyurethane SMPs reinforced with CB alone show only 30% recovery. The interactions of anisotropic CNT with crystallizing polyurethane switching segments result in highly accurate shape recovery. This shows that CNTs are superior to CB as reinforcements for SMPCs in terms of shape recovery.

Due to the high thermal conductivity (nearly 3000 W/mK) of CNTs, they are good candidates for heat conduction inside SMPCs that can be thermally activated. The molecular dynamics studies of polymer nanocomposites (aligned multi-walled nanotubes [MWNTs]/single-walled carbon nanotubes [SWCNTs]) help to better assess the thermal properties. The out-of-plane and interlaminar properties of advanced composite structures are enhanced by the use of these materials.

A new light-induced SMP has been reported (Zhang et al., 2008) in which infrared light is used on surface-modified optical fiber–reinforced polymer composites (the cladding of which is removed by treatment with sodium hydroxide), as depicted in Figure 7.5a,b. The embedded network of light-transmitting optical fibers efficiently transfers heat to the matrix to activate shape memory effect (SME) and achieve faster shape memory.

This method is different from mechanisms such as Joule heating, direct heating, or magnetically induced shape changing, as practiced with SMPCs,

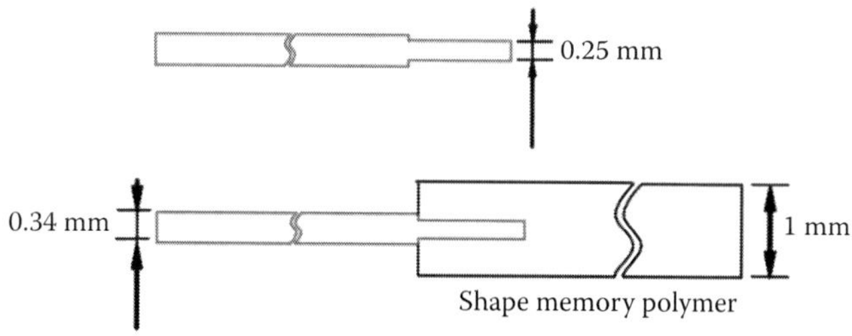

(a)

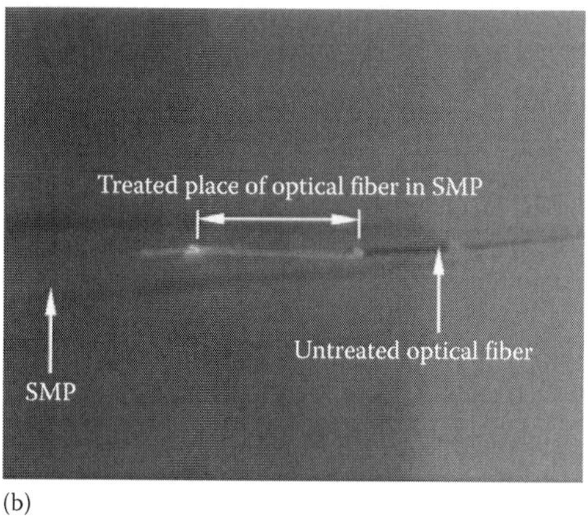

(b)

FIGURE 7.5
(a) Treated optical fibers embedded in the SMP. (b) Laser light of wavelength 632.8 nm coupled with the end of the optical fiber outside the SMP. (From Zhang, D. et al., 2008, Infrared laser-activated shape memory actuators, *Earth and Space*, 40988: 1–7, with permission from ASCE.)

which could open up the technology to newer applications. During the macroscopic shape recovery process, the shape recovery performance of the SMPC is decided not only by the shape memory behavior of the SMP but also by the microstructural deformation mechanism of the fibers and the SMP.

The tensile stiffness of carbon fiber is much higher than its compressive stiffness and the stiffness of SMP matrices. Thus, we assume that, upon applying a bending force to the SMPC specimen, the neutral axis of the specimen moves from the middle plane toward the outer surface, where the fibers are in a state of tensile stress. SMPC materials are similar to traditional fiber-reinforced composites, except for the use of a thermoset styrene-based shape memory resin that withstands much higher packaging strains than traditional composites without damage to the fibers or the resin.

The high strain capacity allows SMPC component designs that can be packaged more compactly than designs made with other materials. In order to achieve a high package strain and avoid fiber failure in storage, fiber microbuckling is needed. With microbuckling, SMPC materials are suited for use in deployable space structure components because of their high strain-to-failure capability. Note that the buckling and delamination in composites will unavoidably reduce the mechanical properties of the SMPC. Due to the effects of microbuckling and shifts in the neutral-strain plane, fiber-reinforced SMPC laminates can withstand much larger compressive strains than traditional hard-resin composites.

The ability to undergo compressive strains is provided by the microbuckling in the bending deformation of SMPCs, which is an elastic and recoverable response. During microbuckling, the fiber direction stiffness is significantly decreased, and thus the fiber carries less compression load and the neutral-strain surface shifts toward the outside of the bend. Another feature is that, during microbuckling, composites can withstand higher compressive strains without failure. The failure strain in SMPCs is attributed to fiber failure strains, as the system is made up of brittle fibers embedded in a ductile matrix, as is the case with any composite, macroscopically. The flexural strength of a material is decided by the tensile strength rather than its compressive strength, as compressive strength in the direction perpendicular to the reinforcement's plane is several times higher than the tensile strength in the transverse direction.

7.4 Elastic Memory Composites

As an ideal combination of properties for deployable space structures, a new class of material known as *elastic memory composite* (EMC) has been derived from shape memory resins. EMC materials retain the structural properties of traditional fiber-reinforced composites while also functioning as SMMs. Thus, structurally efficient, controllable, deployable space structures and deployment mechanisms can be created with no mechanical moving parts. Due to their processability, epoxy matrices and their composites provide a good means of EMC fabrication, and most of the reported EMC deployable devices/prototypes are made from epoxy- or cyanate ester–based shape memory composites. Epoxy resin–based SMPCs are interesting from the point of view of the ease of their structural design and processing them into EMCs. EMCs have been reported to have good mechanical properties that facilitate different structural applications such as lightweight deployable space structures (Liu et al., 2008). Most of the research work on epoxy-based EMCs is concentrated on the high vacuum, ionizing radiation, UV radiation, atomic oxygen, and debris in space.

The maximum stress levels are higher in the case of high-elasticity SMPs compared with the low-elasticity SMPs for the same weight fraction of fiber. Low and high elasticity are delineated based on the storage modulus values. Generally, the elastic modulus will have a direct impact on the stress levels of normal composite materials. (Qing et al. 2013) observed that specimens with low-elasticity SMPs have superior resistance to cyclic loading in comparison with specimens with high-elasticity SMPs and recovery ratios, which may be due to the contribution of the elastic phase more than the glassy phase.

Epoxy laminates reinforced by carbon fibers, glass fibers, and aramid fibers have been studied by researchers across the world, who have demonstrated the effect of microfibers/fabrics on the stiffness of shape memory epoxy resins (SMEPs), which in turn favor better EMCs (Santhosh Kumar et al., 2013). In a 350 h test, the specimen repeatedly showed 95% shape recovery. The consistent maximum shape recovery shows the worthiness of carbon fibers as reinforcement and epoxy as resin for SMPCs.

Various weight fractions of organic montmorillonite (OMMT) were added to SMEPs and 3% added weight showed improvements in the composites' toughness, tensile strength, and shape recovery time, while the shape recovery ratio was unaffected. There was also an increase in T_g in OMMT-modified SMEPs compared with neat SMEPs.

Even though theoretical calculations and laboratory experimentations prove the mechanism and the properties, the real challenge is to synthesize and realize the experimental/predicted figures.

7.5 Synthesis of SMPCs

The methods employed for the synthesis of SMPCs are the same as those for general composite materials, such as hand layups, spray techniques, filament winding, resin transfer molding, tape-laying and fiber placement systems, autoclave-based methods, film stacking, pultrusion injection molding, sheet molding, and so on. Detailed explanations of this can be found in any standard composite materials textbook.

Application-oriented studies emphasize the synthesis of SMPs with good processability and excellent shape recovery characteristics for deployable composite systems. To synthesize SMPCs for specific applications, the necessary parameters to be addressed are shape recovery, stiffness, strength, and recovery time. At the initial stage, SMPs are designed with these properties, and proper fiber reinforcements are used as matrices to create SMPCs.

The thermomechanical design of SMPCs before synthesis is significant and is done by modifying the molecular structure of the polymer or adding functionalized fillers into the polymer as a multiphase composite matrix. EMCs from SMPCs can be of different types, such as particle or fiber reinforced. To synthesize SMPCs, the fillers used are specific to the characteristics intended for the final application: CB, CNTs, carbon nanofibers, SiC, Ni, Fe_3O_4, clay, and short or continuous fibers. Due to the properties of the fillers/fibers, SMPCs develop high electrical conductivity, magnetic-responsive performance, and high stiffness at the micro scale. Electrically conductive CB, CNTs, nickel powders, and chopped carbon fibers are incorporated into SMPs to develop electroactive SMPCs, which are discussed in detail in the following section. Continuous fiber-reinforced SMPCs show significant improvements in strength, stiffness, and resistance to strain relaxation and creep, thereby providing better mechanical properties. It is interesting to note that, in addition to carbon fiber–reinforced SMPCs, glass fiber– and aramid fiber–reinforced SMPCs are also being studied and developed for the various features associated with the fibers and fillers.

Studies on the synthesis of thermoresponsive low- and high-transition temperature SMPs have made use of the cross-linking reactions of copper (I)-catalyzed alkyne-azide cycloaddition (CuAAC) (Ragin Ramdas et al., 2015). Thus, a polymer with dual-trigger temperatures (83°C and 113°C) was synthesized by *click polymerization*.

EMC structures are lighter, have higher specific strength, and have lower coefficients of thermal expansion than SMAs, which are desirable properties for space-deployable systems and are fabricated using traditional composite fabrication methods. For deployable space structures, an EMC can be heated to its T_g (after it is cured), where it becomes flexible and can be tightly packed (stoved). The flexibility of the SMPC material at the folding temperature (T_g) heavily depends on both the resin and fiber properties, as discussed in Section 7.4. Epoxy resins are ideal for processing composites (reemphasizing the qualification of SMEPs for EMCs, as discussed earlier). Carbon fabric–reinforced epoxy resins have been studied for the processing of EMCs, and their thermomechanical and shape memory properties have been evaluated and correlated to epoxy resin compositions (Liu et al., 2008; Dyana Merline and Reghunadhan Nair, 2012).

The synthesis, processing, and shape memory properties of EMCs prepared from epoxy-anhydride resin and the various properties of reinforcements have been studied. The EMC of the epoxy-anhydride system is prepared by using epoxy resin, carboxy-telechelic poly(tetramethylene oxide) (PTMO), and pyromellitic dianhydride (PMDA). The resin mixture (with various weight fractions in each case) is dissolved in 40% wt acetone at room temperature and then 2 mm-thick prepregs are stacked and molded in a hydraulic press and cured by slowly heating them to their cure temperature (determined by DSC, Fourier transform infrared [FTIR] spectroscopy, and rheology). Shape recovery is independent of the weight percentage of components and remains constant at 97%, but the shape recovery time increases with the carbon content. Thus, it can be inferred that the shape memory property is dictated by the resin and the reinforcements restrain the shape recovery. Hence, if sudden deployment is required, the carbon content may be kept low to optimize shape memory properties other than recovery time, and for applications such as space deployment systems and hinges, a slow recovery is recommended to avoid damage to the structure arising from sudden deployment.

Diels–Alder chemistry is one of the most popular methods for synthesizing SMPs, the shape memory of which can be erased by heating them above 160°C. This is due to the *retro Diels–Alder reaction*, which allows us to think about recyclable SMPs. The three hallmark click polymerizations are Diels–Alder, CuAAC, and thiol-ene, with which thermally stable, trigger temperature–tunable, and self-healable SMPs have been synthesized.

The study of epoxy resin systems has been carried out across the world, with varying weight fractions of resin and reinforcements, different conformations of resin chemistry, and different reinforcement fibers. On synthesis, a composite with a carbon–resin ratio of 50:50 diepoxy system exhibited a maximum glass transition value of 119°C (with maximum shape recovery of 97%). The epoxy Novalac system exhibited a low glass transition value of 54°C, and the tris epoxy system exhibited a glass transition of 100°C, respectively. The maximum value for the glass modulus was observed for the 60:40 resin/reinforcement composition. As in previous sections, the large difference in moduli between the glassy and rubbery states results in a superior SMPC. Shape recovery time increased with carbon content, which establishes the need to optimize the carbon weight fraction. The recovery time was directly proportional to the shape memory resin content of the composite, which indicates the role of the matrix.

The conditions of polymerization and the monomers involved decide the molecular weight of the product, which has an impact on the shape memory properties as well as on T_g. As discussed earlier, a higher T_g is mainly associated with a higher degree of chain entanglement (cross-linking). Molecular weight (Mn) is a function of the chain length and entanglement. Thus, the highest Mn gives a better T_g value than that of the lower Mn polymer. Poly(bisphenyl dianhydride-p-phenylenediamine) (BPDA/PDA) polyimide was used with unidirectional CNT to synthesize an SMPC, and a molecular weight–T_g relation was established.

Another method of composite fabrication using shape memory resin and carbon fiber reinforcement is a winding spray technique that can apply the

FIGURE 7.6
(a) A view of the SEM morphology of a unidirectional MWNT composite with polyimide resin at 10 μm magnification. (b) Stress–strain plot of an MWNT/BPDA–PDA composite vs. pure BPDA–PDA polyimide. (Reprinted from Jiang, Q. et al., 2014, Mechanical, electrical and thermal properties of aligned carbon nanotube/polyimide composites, *Composites, Part B: Engineering*, 56: 408–412, with permission from Elsevier.)

polymer as resin while the CNT ribbons are wound onto a rotating mandrel. This method, equivalent to filament winding in composite manufacturing, is effective for producing CNT composites with high-volume fractions and excellent mechanical, electrical, and thermal properties, without any threat of the delamination/debonding of laminar composites, as is generally experienced.

Polymer composites with SME are synthesized by winding (at a speed of 18 mm/s) the MWNT sheets onto a rotating cylindrical spool of polytetrafluoroethylene (PTFE). A polyimide precursor (polyamic acid) was sprayed onto the MWNT sheets layer by layer during winding. An SEM image of (unidirectional) MWNTs is shown in Figure 7.6a for better understanding of the MWNT–polyimide SMPC.

Figure 7.6b establishes that the mechanical properties of the SMPC are enhanced compared with the properties of the basic matrix material (SMP). An obvious improvement in modulus (53.73 ± 3.29 GPa, almost 12 times larger than that of pure polyimide) is found in the case of the composite with MWNT winding.

7.6 Thermal and Electrical Properties of SMPCs

The impact of CNT density on the thermal conductivity of nanocomposites consisting of mechanically densified, aligned MWNT arrays was investigated by different research groups in studies of polyimide SMPs (Xiao et al., 2016; Ragin Ramdas et al., 2015; Jiang et al., 2014). Aligned MWCNT arrays can potentially provide more direct thermal conductivity

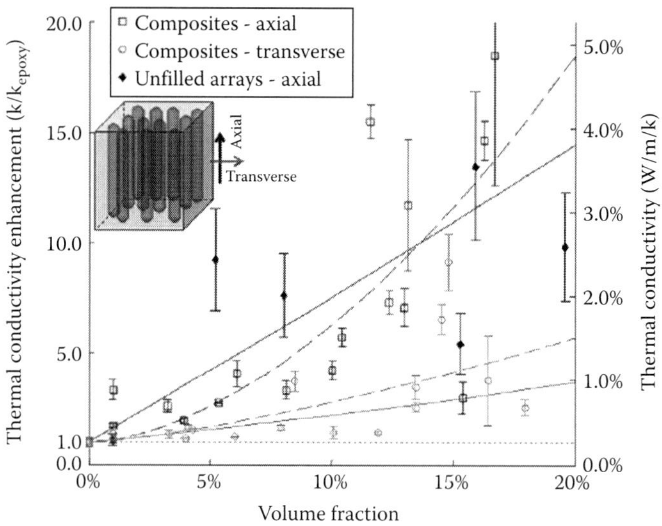

FIGURE 7.7
The axial (□ or indicated by a dark gray line) and transverse (○ or indicated as a light gray line) thermal conductivity of CNT nanocomposites plotted against volume fraction. The dashed line at the bottom indicates the thermal conductivity of the epoxy. (Reprinted from Marconnet, A. M. et al., 2011, *ACS Nano*, 5(6), 4818–4825, with permission from the American Chemical Society.)

pathways across the entire composite thickness (Figure 7.7). SMPCs with controlled CNT content allow correlations between the density and CTE of the composite, which can be a reference for designing nanocomposites for specific applications. The inset picture on the top left of Figure 7.7 with fibers aligned in resin shows the sample configuration used for the study. The plot for the rising curves can be interpreted as an enhancement of conductivity with an increasing volume fraction of CNTs in the bulk.

The axial thermal conductivity averages to a plot rising from 0.4 toward 4.8 W/mK (shown by the dashed lines). The transverse thermal conductivity change is very gradual with respect to the change in volume fraction as it crawls from 0.4 to 1.5 W/mK as the volume fraction approaches 20%. For comparison, the composite matrix without CNT filler is shown as an unfilled array with no particular trend.

The aspect ratio of MWNTs has a great influence on the electrical percolation network. A higher aspect ratio means the electrical percolation network forms at a lower content of MWNTs (even when the weight percentage of MWNTs is below 1%). Similar to SWCNTs, the volume fraction of MWNTs in the matrix is also evaluated (Figure 7.8). Plot (a) shows the reduction in electrical resistivity of the system due to the extremely good conductivity of carbon, and plot (b) shows a visualization of the optimization curve of the volume fraction for the best thermal conductivity.

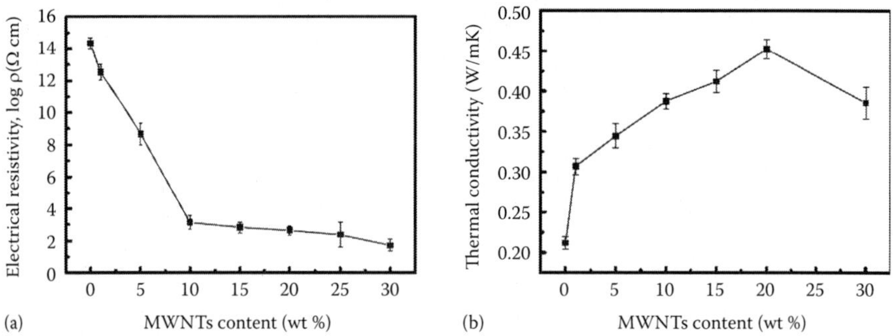

FIGURE 7.8
(a) Volume electrical resistivity. (b) Thermal conductivity of polyvinyl alcohol/MWNT composites with different contents of MWNT loading. (Reprinted from Du, F. P. et al., 2015, Electroactive shape memory polymer based on optimized multi-walled carbon nanotubes/polyvinyl alcohol nanocomposites, *Composites, Part B: Engineering*, 68: 170–175, with permission from Elsevier.)

Thus, it is clear that CNTs as fillers improve the thermal conductivity of polymer matrices. As with any matrix reinforcement composite system, interfacial thermal resistance is one of the critical issues that hinders the improvement of thermal conductivity for carbon-reinforced SMPCs due to phonons (a quantum of energy or a quasiparticle associated with a compression wave such as sound or the vibration of a crystal lattice) or a mismatch of vibration energy and severe phonon scattering at the interface of the CNTs and the polymer. To solve this issue, the CNT surfaces are modified, so that the compatibility and interfacial thermal conduction can be improved by strong interactions between the polymer and the CNTs.

As an extension of the preceding two observations about strength as a criterion, between the MWNT and SWNT studies, nonuniform axial deformations inside MWNTs cause a reduction in strength. Thus, SWNTs may be preferred as reinforcement, considering the strength aspect, as MWNTs and SWNTs behave similarly in terms of their electrical and thermal aspects.

CB is chosen by many researchers as filler material for experimentation with SMPCs to compare the results of studies done with composites containing nanoparticles with that blended with microsized conductive fillers. Thus, the focus of filler in SMPCs narrows itself down to nanosized particles.

Hence, it is clear that the parameters that drive design and synthesis are specific to the applications of SMPCs, and the demands of the application environment become an impetus for the research work associated with the improvement of each characteristic of such materials.

7.7 Applications of SMPCs

As discussed in Chapter 2 on the applications of SMAs, SMPCs also spread widely across many fields of engineering, replacing conventional complex mechanisms. SMPs (in SMPCs) in many practical applications often undergo large three-dimensional deformations influenced by the filler, the matrix material, and the triggering stimuli. Operating temperature is also a factor that affects the work of such materials in sensitive/precise operating conditions.

SMPCs are in their most raw and simple form as heat-shrinkable polymer tubing and film, which have routine applications such as safety tags and self-deploying chairs. Biological applications are at the forefront of the research into SMPCs as surgical tools, and products have been simplified due to the introduction of SMPs in place of complicated mechanisms. SMA or SMP cardiovascular stents are deployed inside the blood vessel during angioplasty. The deployment mechanism is the thermal stimulation of the SMM, and different strategies have been developed to heat the polymer inside the human body. The most common techniques are the use of IR lasers and magnetic nanoparticles to precisely open the stent at the right location. Other than polymer vascular stents, SMPCs are used for orthopedics, endoscopic surgery, orthodontics, kidney dialysis, photodynamic light therapy, aneurysm therapy or neuroprosthetics, drug delivery systems, smart surgical sutures, and laser-activated SMP microactuators to remove a clot in a blood vessel. Two-way SMPs facilitate reversible actuation in devices such as artificial muscles made from the EMCs of SMPCs.

Internal clinical sutures and stitches are a challenge in endoscopic surgeries near an incision or an open lumen. They can be replaced using SMPC surgical suture threads that are trained for a temporary low-tension configuration that upon stimulation can memorize their original tightness (Figure 7.9). Sutures in cryosurgery use the same technique for wound closure, as the body temperature acts as the stimulation for shape recovery.

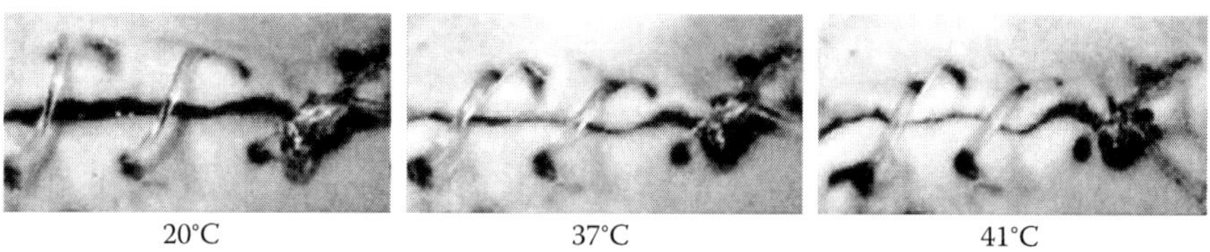

FIGURE 7.9
SMP suture for wound closure responds to body temperature at room temperature and tightens. (Reprinted from Leng, J. et al., 2011, Shape-memory polymers and their composites: Stimulus methods and applications, *Progress in Material Science*, 56(7): 1077–1135, with permission from Elsevier.)

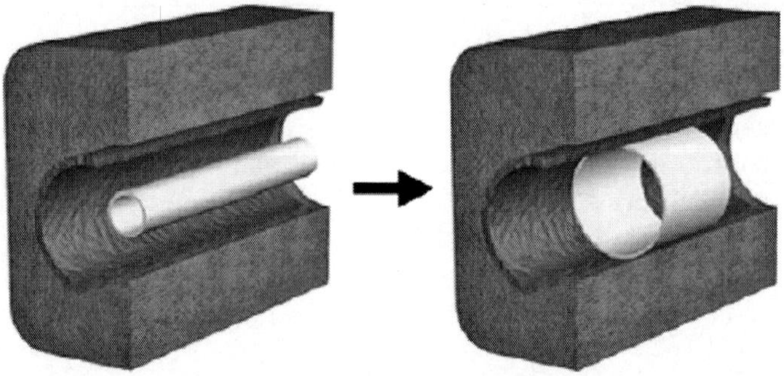

FIGURE 7.10
Analogy to illustrate the use of SMP stents in angioplasty. (Reprinted from Leng, J. et al., 2011, Shape-memory polymers and their composites: Stimulus methods and applications, *Progress in Material Science*, 56(7): 1077–1135, with permission from Elsevier.)

The same concept can be used for the deployment of different clinical devices implanted in the human body (Figure 7.10). SMPC stents used as drug delivery systems/probes to kill cancerous cells help to significantly reduce restenosis and thrombosis. They are inserted into body in a temporary trained shape and expanded at the target site. This may or may not be removed at a later point, depending on the biocompatibility and the requirements of such a system inside the body.

Figure 7.11 illustrates thermo-/moisture-responsive microdevice delivery inside a cancer cell for clinical purposes. The delivery device enters the body in a temporary shape and traverses to the affected cell, where it is stimulated by suitable activation methods and regains its original shape to execute its purpose (drug delivery/root removal) at the cancer cell. SMPs have micro-/nanoscale actuation capabilities for microelectromechanical and nanoelectromechanical systems, micro-/nanopatterning, biomedical devices, and so

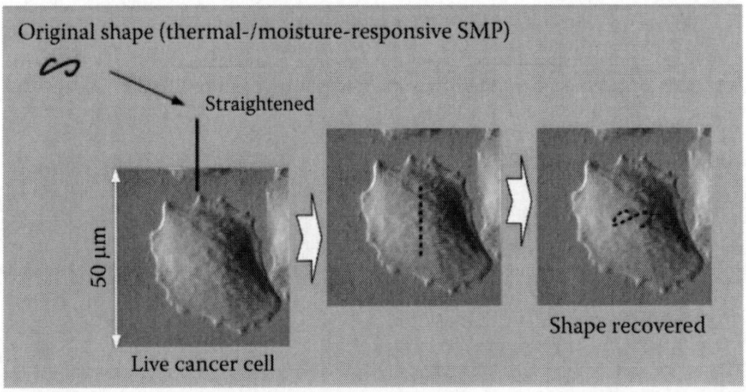

FIGURE 7.11
Concept of micro-/nanodevice delivery to a living cell. (Reprinted from Leng, J. et al., 2011, Shape-memory polymers and their composites: Stimulus methods and applications, *Progress in Material Science*, 56(7): 1077–1135, with permission from Elsevier.)

on. Besides biomedical applications, SMPCs find many uses in space-deployable structures such as hinges, trusses, mirrors and reflectors, morphing skins used for folding or variable camber wings in aircraft, SMPC textiles, automobile actuators, and self-healing systems.

One decisive advantage of SMPCs when considering their aerospace applications is their low density. It is noteworthy that the latest Boeing 787 Dreamliner uses more than 50% polymer composites in the primary structure, which removes a large amount of weight. In spacecraft structures, SMPCs are influential due to their low-cost, self-deployable, less complex structures, such as solar arrays, solar sails, sunshields, or radar antennas.

Application-specific research has led to the development of cold hibernated elastic memory (CHEM) technology for spacecraft structures. Witold Sokolowski invented CHEM materials to be used in self-deployable and rigidizable structures. CHEM foams are self-deployable materials that use elastic recovery and are under development at NASA's Jet Propulsion Laboratory (JPL) and Mitsubishi Heavy Industries (MHI). The advantage of the CHEM concept is the possibility of realizing lightweight, compressible structures that memorize their original shape even after long-term compressed storage. This is very similar to SMPs but involves the critical softening temperature of storage (Figure 7.12). In CHEM, an object constructed of thermoplastic polymer foam is heated above the SMP's softening temperature (T_g) and compacted to a desired reduced volume; the temperature then is reduced below T_g for storage. It utilizes SMP foam structures, or sandwich structures made of SMP foam cores and polymeric composite skins, and is proposed for space-bound structural applications by Sokolowski and Hayashi (2003).

An interesting comparison of the weight of the current inflatable sun shield of the International Space Station (ISS) with a CHEM-based system demonstrates the obvious advantages of CHEM materials for space-deployable structures. The actual weight of the ISS's inflatable sun shield is approximately 105 kg, while an alternative CHEM module would weigh less by an order of 10. Thus, very lightweight space structures help to gain payload for the launch vehicle as well as create less complex systems for the spacecraft (e.g., deployable panels [discussed later in the chapter], solar arrays, and reflector antennas).

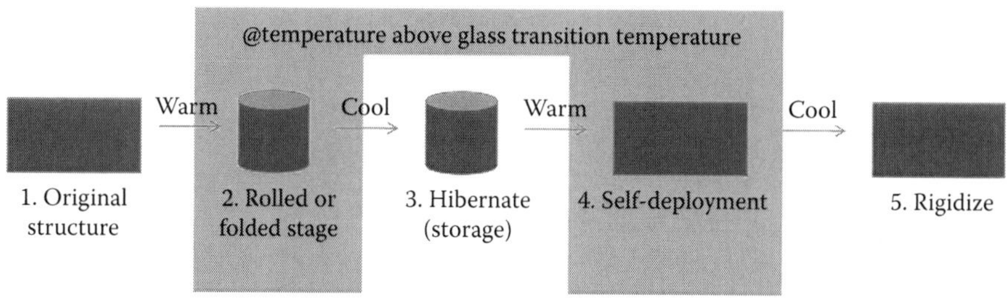

FIGURE 7.12
Representation of the concept of CHEM.

Since self-healing structures significantly extend the lifetime and utility of smart structures, the self-healing of structural damage has remarkable significance for researchers. It is interesting to note that SME and self-healing techniques are opposites in their working principles. Upon sensing damage, the monitoring system identifies the location coordinates and estimates the magnitude of damage. Smart sensors and actuation signals to the controlling system, and the local heating of the location induces the SMP to return to its originally designed shape and thus heal the damage. To realize conventional self-healing approaches such as microcapsules and hollow fibers in SMPCs, it demands a good amount of field research. The *reversible plasticity shape memory effect* (RPSME) allows the plastic deformation of a specific area in the system to fill the gaps and rectify the damages.

Microgravity does not have any effect on an SMPC laminate structure's ability to recover its shape. Exoatmospheric/outer-space operations impose limits on the heating system design due to the differences in heat transfer on Earth and in orbit. This requires an exclusive, thermally isolated system with a cooling mechanism to facilitate controlled heat transfer in remote actuations. The complex geometry of deployment will be affected for in-orbit structures, and hence new structural composite actuators with relatively simple configurations are designed from complex geometry (Figure 7.13).

The need for deployable structures in space applications is significant when the physical dimensions of the spacecraft's operational configuration are much larger than the launch vehicle firing dimensions due to the functional aspects.

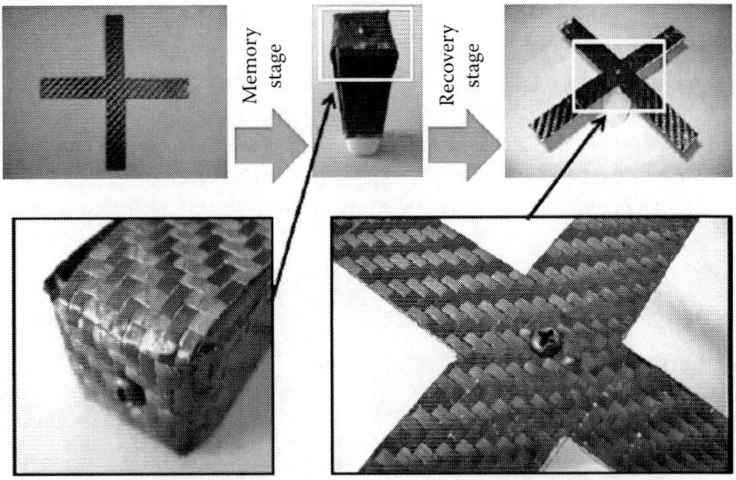

FIGURE 7.13
Memory and recovery stages of the SMC cross. (Reprinted from Santo, L. et al., 2014, Shape memory composites for self-deployable structures in aerospace applications, *Procedia Engineering*, 88: 42–47, with permission from Elsevier.)

As discussed, fiber-reinforced SMPCs show high strain-to-failure capabilities under flexure deformation, due to the microbuckling phenomenon. This makes SMPCs suitable for use in deployable space structure components. Fiber microbuckling ensures high packaging strain and avoids fiber failure in SMPCs, but material nonlinearity is an additional constraint. The drawbacks of traditional deployment devices/mechanisms, such as the complex assembly processes, massive mechanisms, large volumes, and undesired effects during deployment, are overcome using SMPCs. One such concept is

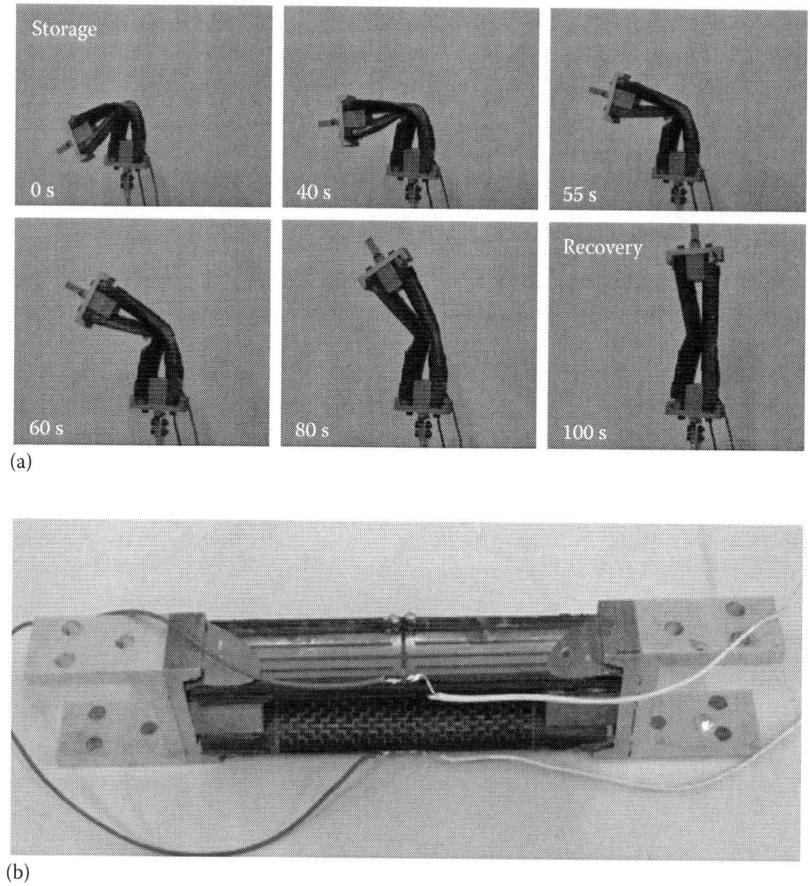

FIGURE 7.14
Shape memory hinge. The left-hand figure shows the various stages in the deployment process of a hinge made of SMPC activated by thermal stimulation initiated by Joule heating, which by end of 100 s regains its original shape and proves the possibility of stressed recovery of SMPCs. The right-hand figure shows the prototype SMPC hinge along with the conductive part that heats up for shape recovery. (Reprinted from Leng, J. et al., 2011, Shape-memory polymers and their composites: Stimulus methods and applications, *Progress in Material Science*, 56(7): 1077–1135, with permission from Elsevier.)

an SMPC hinge, as shown in Figure 7.14, that opens up the structure without shock, as in pyrosystems used currently.

Demonstrations of deployment in simulated microgravity conditions prove the feasibility of the hinge mechanism for space structures. The hinge is a styrene-based SMPC made for the actuation of deployment for solar arrays. Leng et al. (2011) focused on the influence of fiber reinforcement on the glass transition of the SMPC, the influence of temperature on the bending modulus of the SMPC, the deployment properties upon bending the SMPC, and the deployment sequence of a prototype solar array actuated by the SMPC hinge. The temperature is achieved in the system by Joule heating, which is done by converting the energy from an electrical current into heat energy. Upon applying an electric field, a difference in electric pressure is created, commonly referred to as *potential*, and electrons from atoms are set into motion due to the potential created. Thus, each time an electron bounces off an atom in a material, it creates kinetic energy in the system. This energy, in the form of vibration, is transferred to the adjacent atoms, and the overall combination of different vibrations is given away as heat.

To heat the system, 20 V is applied to the resistor embedded in each circular laminate (Figure 7.15 illustrates the deployment of the solar array prototype with the SMPC hinge). The spacecraft needs lightweight, reliable, and cost-effective mechanisms for the deployment of radiators, solar arrays, or other devices, as the launch space inside a payload is limited in current launch vehicle scenarios and needs shock-proof deployment in orbit. Figure 7.15 illustrates the testing of the prototype hinge system, simulating a zero-G condition in a laboratory.

An EMC lightweight extendable boom was designed and realized by Composite Technology Development (CTD) for use in a microsatellite. Elastic memory is being utilized to recover the original shape from a collapsed configuration during launch. Three longitudinal members or *longerons* act as the directive for the EMC component to deploy itself as a memory-activated boom upon receiving stimuli.

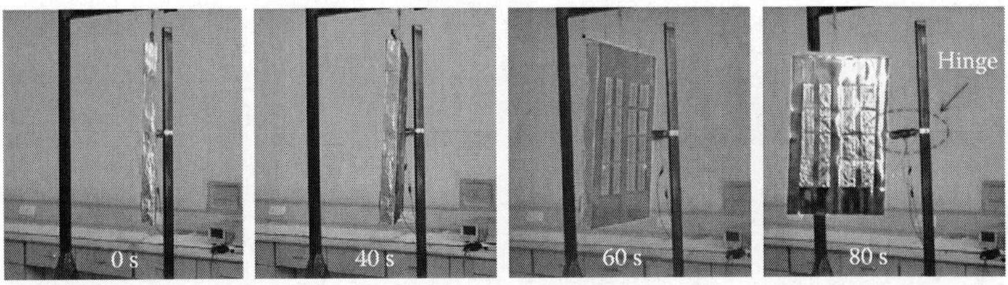

FIGURE 7.15
Solar array prototype actuated by an SMPC hinge deployment demonstration. A tension wire during operation creates a zero-weight system that simulates a zero-G environment. (Reprinted from Leng, J. et al., 2011, Shape-memory polymers and their composites: Stimulus methods and applications, *Progress in Material Science*, 56(7): 1077–1135, with permission from Elsevier.)

Liu et al. (2014) report the use of a simple solar array with very few moving parts and low costs. The array was designed by CTD as a result of their research into SMPCs for space applications, in a joint effort with the NASA JPL. The fabrication of the deployable panels of the Intelligent Nanosat Operations Satellite (DINO Sat) is a pioneering step toward EMC hinge–based solar array deployment in orbit.

As addressed earlier, conventional pyrotechnics, which can even cause the operational failure of spacecraft, as history testifies, are being replaced with intelligent systems. The concept of accommodating larger structures that can be stowed inside the confined heat shield space supplements the maximum use of the launch space inside a payload with lightweight SMP materials. This makes optimal use of payload space and can positively affect the launch cost per rocket.

Numerous deployable antenna structures in various configurations are being designed that make use of smart materials. One such lightweight system is depicted in Figure 7.16.

In this reflector, the inner and outer circumferential element is made of EMC, and actuating a deployment sequence will open up the antenna. It occupies less space in the stowed configuration and unfurls to the required aperture on orbit.

SMPC materials are used to fabricate the central part of the antenna, to enable a large deformation along with the supporting structures. Self-healing helps to minimize errors as the antenna can adapt to severe space environments.

Thus, the advantages of SMPCs (variable moduli, large recovery rates and forces, high stability, low weight, and reliability) are being exploited in these space structures to make them lighter than the structures currently used in space. Simultaneous and sequential deployment schemes are being explored, of which the former is a less controlled, less complex, and lower mass deployment mechanism, and the latter uses less power as it is required to heat fewer joints at a time.

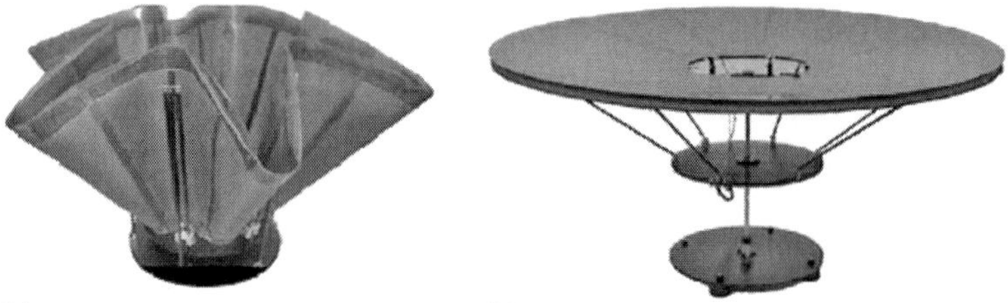

(a) (b)

FIGURE 7.16
SMPC reflector: stowed and recovered shapes. (Reprinted from Leng, J. et al., 2011, Shape-memory polymers and their composites: Stimulus methods and applications, *Progress in Material Science*, 56(7): 1077–1135, with permission from Elsevier.)

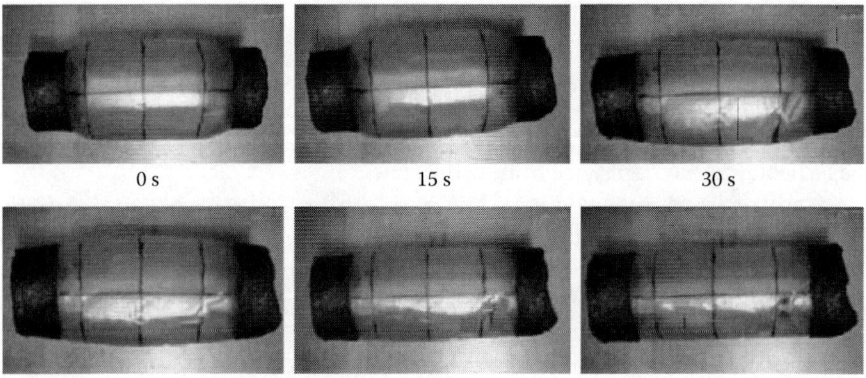

FIGURE 7.17
The shape recovery process of a bottle-shaped smart mandrel. (Reprinted from Zhang, L. et al., 2014, Analysis and design of smart mandrels using shape memory polymers, *Composites, Part B: Engineering*, 59: 230–237, with permission from Elsevier.)

Another innovation in SMPC applications is the development of smart mandrels, whereas the traditional water-soluble, elastomeric mandrels, and multipiece metal mandrels have inherent disadvantages; they are time-consuming, expensive, and difficulty to remove. Reusable SMPC smart mandrels can retain their temporary shape and recover their original shape with certain external stimuli. Good shape fixity confirms the accuracy of the mandrel and the finish of the products, and good shape recovery is required for successful extraction from the curing composite product. A demonstration of the operation of a bottle-shaped mandrel at 90°C (T_g 63°C) is shown in Figure 7.17 as a sequence of deployment that lasts slightly more than a minute.

Similar to the CTD research groups mentioned previously, the Cornerstone Research Group (CRG) has developed a lightweight deployable ground-based mirror with an SMPC substrate and a coated reflective side in a sandwich construction. The support structure is a honeycomb core fabricated with SMPs topped by a mirror surface. Carbon nanofibers are included for activation and thermal conductivity enhancement.

Figure 7.18a depicts a model of the reflective mirror, which can alter the curvature with suitable stimuli. This is made of polymeric materials (the cross-section shown in Figure 7.18b) with a stiffened honeycomb base. The microspheres allow the relative movement of layers during shape changing, as the fabric/ribbon inside the penultimate layer connects to the carbon fiber layer and makes the conductive path continuous.

Different shapes correspond to a trade-off between beneficial characteristics, such as speed, low energy consumption, and maneuverability. Since SMPs have the ability to change their elastic moduli (as discussed

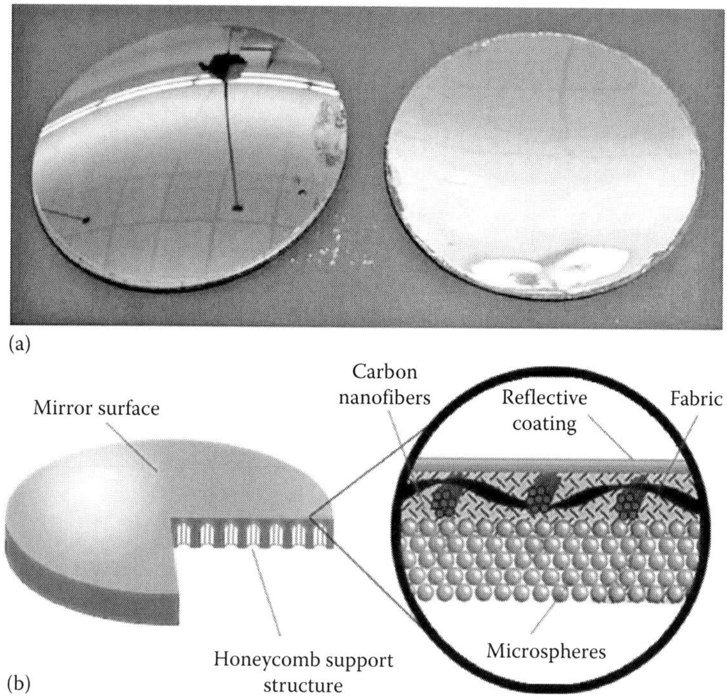

FIGURE 7.18
Reflective SMP mirror concept. (a) Synthesized SMP mirror. (b) Concept of SME utilization for the polymeric material used for synthesis. (Reprinted from Leng, J. et al., 2011, Shape-memory polymers and their composites: Stimulus methods and applications, *Progress in Material Science*, 56(7): 1077–1135, with permission from Elsevier.)

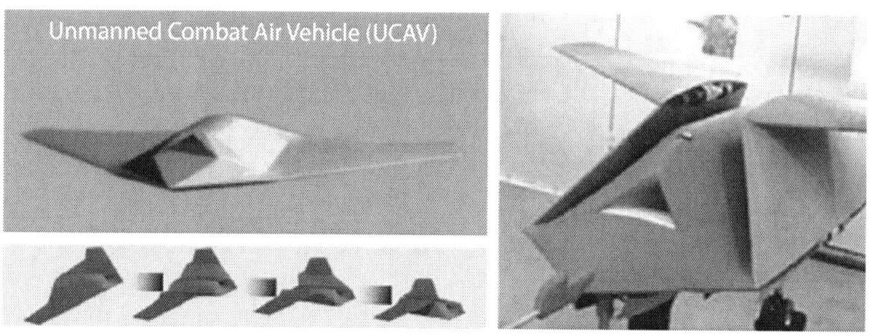

FIGURE 7.19
Morphing wing developed for an unmanned combat air vehicle (UCAV) and tested by Lockheed Martin. (Reprinted from Leng, J. et al., 2011, Shape-memory polymers and their composites: Stimulus methods and applications, *Progress in Material Science*, 56(7): 1077–1135, with permission from Elsevier.)

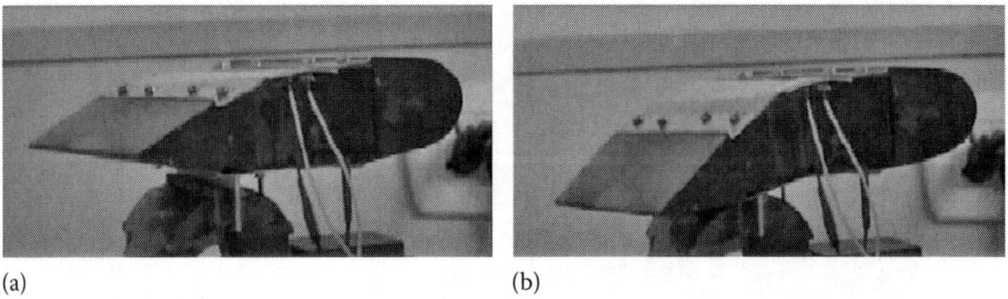

FIGURE 7.20
Variable camber wing with (a) original and (b) morphed configurations. An angular shift of 15° is achieved at the laboratory scale, a clear illustration of the load-bearing capacity of SMPs, which will aid further applications. (Reprinted from Leng, J. et al., 2011, Shape-memory polymers and their composites: Stimulus methods and applications, *Progress in Material Science,* 56(7): 1077–1135, with permission from Elsevier.)

in Section 7.2.1), they could potentially be used in the morphing of a variable camber wing, as shown in Figure 7.19. As per this concept, morphing aircrafts can change wing shape during flight to adapt to the aerodynamic requirements. The concept is obviously bioinspired, from the basic idea that the wings of birds are far from being rigid, once again demonstrating the influence of biomimetics on smart materials and technologies. Lockheed Martin published the adaptive wing drone concept shown in Figure 7.19, which regulates the vehicle speed and aerodynamics based on the morphing technique.

The concept of a morphing variable camber wing comprised of a flexible SMP skin, a metal sheet, and a honeycomb structure has been revolutionary. The metal sheet is used to replace traditional hinges to keep the surface smooth during camber changing. The honeycomb structure provides stiffness for the entire unit, preventing any dimensional changes in the perpendicular in-plane axis, and the flexible SMP skin is covered to create the smooth aerodynamic surface. A test setting at the laboratory scale is depicted in Figure 7.20.

Recovery time has been a concern for robotic applications as less recovery time is advisable, while space deployment of stowed structures can afford for a slow recovery to prevent shock or damage to the sensitive instruments on sudden deployment. Thus, a variable order of recovery speed based on application demands development of materials with different shape change speed based on stimuli and component characteristics of composites in SMPC.

Considering the pace of space exploration, establishing a base outside earth for interplanetary/interstellar exploration is an immediate requirement. This demands a habitat to sustain basic life and avoid damage from exo-Earth radiation. As has been well established from the preceding discussions, smart structures are an obvious solution to these concerns.

Inflatable structures to be used as habitats/platforms for scientific exploration in space are being studied by scientists, and many prototypes and tests are in progress (an inflatable lunar habitat is being studied by NASA as per the published literature [Liu et al., 2014]). The basic concept is to make the supportive structural frames out of an SMPC that deploys upon stimulation. This can help to give composite frames the required stiffness for study platforms in space.

Apart from applications such as mirrors, deployable optic reflectors, trusses, deployable antennas/reflectors, biomedical use, robotics, and so on, as previously explained, research has been initiated into the use of SMPCs in airflow control systems. For high-speed flow research work in long sustainable burning scramjets or air breathing propulsion, the mentioned airflow control system research using SMPCs can provide better alternatives and optimal solutions.

References

Du, F. P., Ye, E. Z., Yang, W., Shen, T. H., Tang, C. Y., Xie, X. L., Zhou, X. P., and Law, W. C., 2015. Electroactive shape memory polymer based on optimized multi-walled carbon nanotubes/polyvinyl alcohol nanocomposites, *Composites Part B: Engineering*, 68: 170–175.

Dyana Merline, J. and Reghunadhan Nair, C. P., 2012. Carbon/epoxy resin based elastic memory composites, *Eurasian Chemico-Technological Journal*, 14(3): 227–232.

Gall, K., 2002. Shape memory polymer nanocomposites, *Acta Materialia*, 50(20): 5115–5126.

Jiang, Q., Wang, X., Zhu, Y., Hui, D., and Qiu, Y., 2014. Mechanical, electrical and thermal properties of aligned carbon nanotube/polyimide composites, *Composites Part B: Engineering*, 56: 408–412.

Leng, J., Lan, X., Liu, Y., and Du, S., 2011. Shape-memory polymers and their composites: Stimulus methods and applications, *Progress in Materials Science*, 56(7): 1077–1135.

Liu, Y., Du, H., Liu, L., and Leng, J., 2014. Shape memory polymers and their composites in aerospace applications: A review, *Smart Materials and Structures*, 23(2): 023001.

Liu, Y., Wang, X., Lan, X., Lv, H., and Leng, J., 2008. Shape memory polymer composite and its application in deployable hinge for space structure, *Proceedings of SPIE: Sensors and Smart Structures Technologies for Civil, Mechanical, and Aerospace Systems*, 6932(2): 10–18.

Marconnet, A. M., Yamamoto, N., Panzer, M. A., Wardle, B. L., and Goodson, K. E., 2011. Thermal conduction in aligned carbon nanotube–polymer nanocomposites with high packing density, *American Chemical Society Nano*, 5(6): 4818–4825.

Mather, P. T., Luo, X., and Rousseau, I. A., 2009. Shape memory polymer research, *Annual Review of Materials Research*, 39: 445–471.

Ragin Ramdas, M., Santhosh Kumar, K. S., and Reghunadhan Nair, C. P., 2015. Synthesis, structure and tunable shape memory properties of polytriazoles: Dual-trigger temperature and repeatable shape recovery, *Journal of Materials Chemistry A*, 3(21): 11596–11606.

Qing, Q. N., Ohki, T., Ohsako, N., and Iwamoto, M., 2013. Thermo-mechanical behavior of smart composites with shape memory polymer, *Proceedings of International Conference on Composite Materials: Smart Materials and Structures*, Beijing, China, ID-1332.

Santhosh Kumar, K. S., Biju, R., and Reghunadhan Nair, C. P., 2013. Progress in shape memory epoxy resins, *Reactive and Functional Polymers*, 73(2): 421–430.

Santo, L., Quadrini, F., Accettura, A., and Villadei, W., 2014. Shape memory composites for self-deployable structures in aerospace applications, *Procedia Engineering*, 88: 42–47.

Sokolowski, W. M. and Hayashi, S., 2003. Applications of cold hibernated elastic memory (CHEM) structures, *Proceedings of SPIE: The International Society for Optical Engineering*, 5056: 534–544.

Xiao, X., Qiu, X., Kong, D., Zhang, W., Liu, Y., and Leng, J., 2016. Optically transparent high temperature shape memory polymers, *Soft Matter*, 12(11): 2894–2900.

Zhang, D., Liu, Y., and Leng, J., 2008. Infrared laser-activated shape memory actuators, *Earth and Space*, 40988: 1–7.

Zhang, L., Du, H., Liu, L., Liu, Y., and Leng, J., 2014. Analysis and design of smart mandrels using shape memory polymers, *Composites Part B: Engineering*, 59: 230–237.

Bibliography

Aguilar, J. O., Bautista-Quijano, J. R., and Avilés, F., 2010. Influence of carbon nanotube clustering on the electrical conductivity of polymer composite films, *Express Polymer Letters*, 4(5): 292–299.

Chaterji, S., Kwon, K., and Park, K., 2005. Smart polymeric gels: Redefining the limits of biomedical devices, *Biophysical Chemistry*, 257(5): 2432–2437.

Chawla, K., 2014. *Composite Materials: Science and Engineering*, New York: Springer.

Gall, K., 2002. Shape memory polymer nanocomposites, *Acta Materialia*, 50(20): 5115–5126.

Gross, K. E., 2008. Mechanical characterization of shape memory polymers to assess candidacy as morphing aircraft skin, postgraduate thesis, University of Pittsburgh, PA.

Hager, M. D., Bode, S., Weber, C., and Schubert, U. S., 2015. Shape memory polymers: Past, present and future developments, *Progress in Polymer Science*, 49: 3–33.

Han, Z. and Fina, A., 2011. Thermal conductivity of carbon nanotubes and their polymer nanocomposites: A review, *Progress in Polymer Science (Oxford)*, 36(7): 914–944.

Hu, J., Zhu, Y., Huang, H., and Lu, J., 2012. Recent advances in shape-memory polymers: Structure, mechanism, functionality, modeling and applications, *Progress in Polymer Science*, 37(12): 1720–1763.

Jiang, Q., Wang, X., Zhu, Y., Hui, D., and Qiu, Y., 2014. Mechanical, electrical and thermal properties of aligned carbon nanotube/polyimide composites, *Composites Part B: Engineering*, 56: 408–412.

Lan, X., Leng, J. S., Liu, Y. J., and Du, S. Y., 2008. Investigate of electrical conductivity of shape-memory polymer filled with carbon black, *Advanced Materials Research*, 47–50: 714–717.

Lan, X., Zhang, R., Liu, Y., and Leng, J., 2011. Fiber reinforced shape-memory polymer composite and its application in deployable hinge in space, *Proceedings of 52nd AIAA/ASME/ASCE/AHS/ASC Structures, Structural Dynamics and Materials Conference*, Denver, CO, pp. 1–10.

Leng, J., Lan, X., Liu, Y., and Du, S., 2011. Shape-memory polymers and their composites: Stimulus methods and applications, *Progress in Materials Science*, 56(7): 1077–1135.

Leng, J., Lu, H., and Du, S., 2008. Conductive shape memory polymer composite technology and its applications in aerospace, *Proceedings of 49th AIAA/ASME/ASCE/AHS/ASC Structures, Structural Dynamics, and Materials Conference; 16th AIAA/ASME/AHS Adaptive Structures Conference*, Schaumburg, IL, pp. 1–8.

Leng, J., Lu, H., Liu, Y., Huang, W. M., and Du, S., 2009. Shape-memory polymers: A class of novel smart materials, *MRS Bulletin*, 34(11): 848–855.

Lin, J., 2006. Shape memory rigidizable inflatable (RI) structures for large space systems applications, *Proceedings of 47th AIAA/ASME/ASCE/AHS/ASC Structures, Structural Dynamics, and Materials Conference*, Newport, RI, May, pp. 1–11.

Liu, Y., Du, H., Liu, L., and Leng, J., 2014. Shape memory polymers and their composites in aerospace applications: A review, *Smart Materials and Structures*, 23(2): 023001.

Ma, L., Zhao, J., Wang, X., Chen, M., Liang, Y., Wang, Z., Yu, Z., and Hedden, R. C., 2015. Effects of carbon black nanoparticles on two-way reversible shape memory in crosslinked polyethylene, *Polymer* (United Kingdom), 56: 490–497.

Merline, J. D., Reghunadhan Nair, C. P., and Ninan, K. N., 2008. Synthesis, characterization, curing and shape memory properties of epoxy-polyether system, *Journal of Macromolecular Science, Part A: Pure and Applied Chemistry*, 45(4): 312–322.

Mohd Jani, J., Leary, M., Subic, A., and Gibson, M. A., 2014. A review of shape memory alloy research, applications and opportunities, *Materials and Design*, 56: 1078–1113.

Pilate, F., Toncheva, A., Dubois, P., and Raquez, J. M., 2016. Shape-memory polymers for multiple applications in the materials world, *European Polymer Journal*, 80: 268–294.

Scalet, G., Auricchio, F., Bonetti, E., Castellani, L., Ferri, D., Pachera, M., and Scavello, F., 2015. An experimental, theoretical and numerical investigation of shape memory polymers, *International Journal of Plasticity*, 67: 127–147.

Shi, Y., 2013. High temperature shape memory polymers and ionomer modified asphalts, doctoral thesis, University of Akron, OH.

Sokolowski, W. M., 2004. US Patent no. 006702976 B2: Cold hibernated elastic memory self-deployable and rigidizable structure and method therefore, March 9.

Taha, O. M. A., Bahrom, M. B., Taha, O. Y., and Aris, M. S., 2015. Experimental study on two way shape memory effect training procedure for Nitinol shape memory alloy, *ARPN Journal of Engineering and Applied Sciences*, 10(17): 7847–7851.

Volk, B., 2005. Characterization of shape memory polymers, Research and Technology Directorate paper, pp. 1–11.

Wei, Z. G., Sandstrom, R., and Miyazaki, S., 1998. Shape-memory materials and hybrid composites for smart systems, Part I: Shape-memory materials, *Journal of Materials Science*, 33(15): 3743–3762.

Wei, Z. G., Tang, C. Y., and Lee, W. B., 1997. Design and fabrication of intelligent composites based on shape memory alloys, *Journal of Materials Processing Technology*, 69(1–3): 68–74.

Winzek, B., Schmitz, S., Rumpf, H., Sterzl, T., Hassdorf, R., Thienhaus, S., Feydt, J., Moske, M., and Quandt, E., 2004. Recent developments in shape memory thin film technology, *Materials Science and Engineering A*, 378(1–2): 40–46.

Wu, X., Huang, W. M., Zhao, Y., Ding, Z., Tang, C., and Zhang, J., 2013. Mechanisms of the shape memory effect in polymeric materials, *Polymers*, 5(4): 1169–1202.

Xiao, X., Kong, D., Qiu, X., Zhang, W., Liu, Y., Zhang, S., Zhang, F., Hu, Y., and Leng, J., 2015. Shape memory polymers with high and low temperature resistant properties, *Scientific Reports*, 5(2): p. 14137.

Xiao, X., Kong, D., Qiu, X., Zhang, W., Zhang, F., Liu, L., Liu, Y., Zhang, S., Hu, Y., and Leng, J., 2015. Shape-memory polymers with adjustable high glass transition temperatures, *Macromolecules*, 48(11): 3582–3589.

Xie, F., Huang, L., Leng, J., and Liu, Y., 2016. Thermoset shape memory polymers and their composites, *Journal of Intelligent Material Systems and Structures*, 27(18): 2433–2455.

Xie, F., Huang, L., Liu, Y., and Leng, J., 2014. Synthesis and characterization of high temperature cyanate-based shape memory polymers with functional polybutadiene/acrylonitrile, *Polymer*, 55(23): 5873–5879.

Yang, D., 2000. Shape memory alloy and smart hybrid composites: Advanced materials for the 21st century, *Materials and Design*, 21(6): 503–505.

Yang, W. G., Lu, H., Huang, W. M., Qi, H. J., Wu, X. L., and Sun, K. Y., 2014. Advanced shape memory technology to reshape product design, manufacturing and recycling, *Polymers*, 6(8): 2287–2308.

Yin, W., Fu, T., Jingcang, L., and Leng, J., 2009. Structural shape sensing for variable camber wing using FBG sensors, *Proceedings of SPIE: The International Society for Optical Engineering*, 7292: 1–11.

Zhang, D., Liu, Y., and Leng, J., 2008. Infrared laser-activated shape memory actuators, *Earth and Space*, 40988: 1–7.

8

High-Temperature Shape Memory Materials

Aerospace applications demand high-temperature sustaining, lightweight materials that can maintain a desirable range of hardness, tensile strength, and resistance to creep, rupture, oxidation, erosion, and phase changes from 300°C to 1200°C under static/cyclic loads in reduced pressure. When selecting such materials, a fine balance is to be maintained between functional demands and manufacturability (Figure 8.1). Structural/functional requirements focus on, for example, the performance of loading conditions (tension, compression, bending, vibration, cyclic, etc.), while manufacturability constraints (geometric limits, environmental aspects, maintainability, etc.) signify the need for reasonable manufacturing/operating costs and durable materials to meet the demand for lightweight, durable, high-performance aerospace materials.

The material selection stage of aerospace structures can be considered a journey through five different steps:

1. Establishing design requirements (functions, constraints, objectives, free variables)
2. Material rejection/selection (based on constraints)
3. Ranking materials in the domain (based on objectives and properties)
4. Researching (based on the constraints and properties of ranked materials)
5. Cultural constraints (expertise in manufacturability)

Figure 8.2 depicts the position of aerospace-grade materials on a plot of strength to service temperature (logarithmic scale). Metals operate in the 20°C–1500°C range and show an improved strength range above 1000 MPa as alloys. Ceramics outrun metals and alloys in operating temperature due to their exceptional stability, which extends from 300°C to 2000°C. They show considerably good strength near 1000 MPa. Technical and nontechnical ceramic bifurcations are based on direct and indirect applications in engineering. Polymers and elastomers, which survive operating temperatures in the range of 40°C–300°C, show a wider range of strength values, from 1 to 100 MPa (polybenzimidazole, approximately 300 MPa, identified for most space applications by NASA). Although polymeric foams exhibit lower strength and offer lower operating temperatures, foams based on poly(phenylquinoxaline), bismalamine tantalum carbide, and niobium sponges are exceptions.

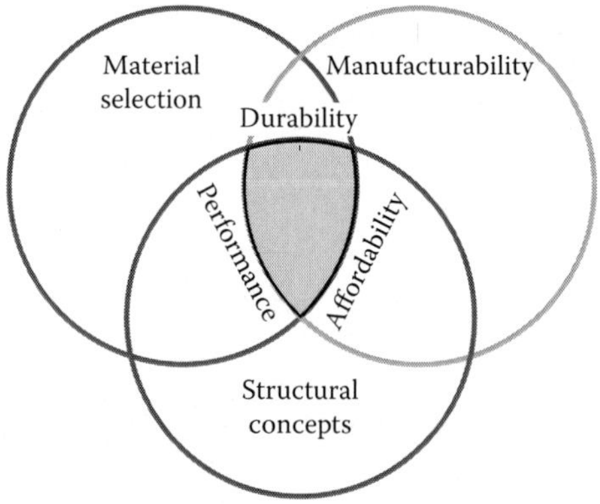

FIGURE 8.1
Engineering utility of materials represented in terms of structural concepts, manufacturability, and material choice. (From Arnold, S. M., 2012, Materials selection for aerospace: Introduction to aerospace materials; NASA/TM-2012-217411, E-18042, NASA Glenn Research Center, Cleveland, OH, 2012.)

High martensitic temperatures (in the case of shape memory alloys [SMAs]) or transition temperatures (in the case of shape memory ceramics/polymers [SMCs/SMPs]) along with acceptable recoverable transformation strain levels, long-term stability, resistance to plastic deformation and creep, and adequate environmental resistance are the qualifying criteria for high-temperature shape memory materials (HTSMMs).

For materials with the shape memory effect (SME), the phase transformation temperature is the critical parameter that decides the trade-off between operating temperature and strength, as discussed in previous chapters, where elastic moduli are evaluated across the transition temperatures.

As shown in Figure 8.2, alloys and carbon composites are ideal materials for a combination of high strength and high operating temperatures. Thus, considering shape memory materials in this category, a good choice of origin will be high-temperature shape memory alloys/polymers (HTSMAs/HTSMPs), which will be discussed in the following sections.

8.1 High-Temperature Shape Memory Alloys

Alloys are molten combinations of two or more metals with enhanced properties in terms of hardness, luster, malleability, fusibility, and ductility, with better electrical and thermal conductivity. Alloys that exhibit excellent mechanical and thermal/electrical properties, improved resistance to thermal creep deformation, and low vulnerability to corrosion and oxidation are termed *high-performance alloys* or *superalloys*. Examples of high-temperature

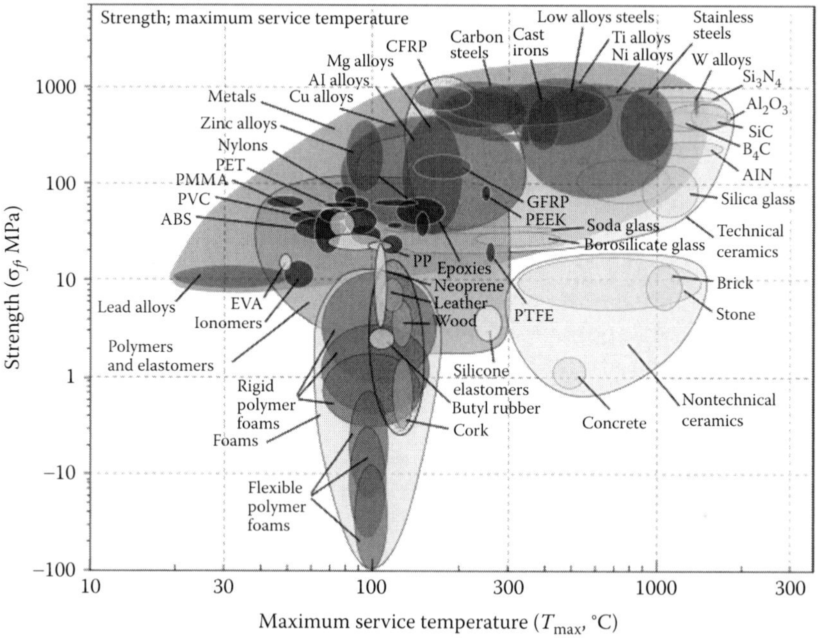

FIGURE 8.2
Strength-to–operating temperature chart of aerospace materials. (From Arnold, S. M., 2012, Materials selection for aerospace: Introduction to aerospace materials; NASA/TM-2012-217411, E-18042, NASA Glenn Research Center, Cleveland, OH, 2012.)

alloys includes Inconel, which is a very strong combination of Ni, Cr, Fe, Mo, Nb, and Ta that can survive temperatures of −40°C to 1800°C, and Hastelloy, which is a combination of Ni, Cr, Fe, and Mo that can survive temperatures up to 1300°C.

As discussed in Chapter 2, the SME process in alloys is a diffusionless solid-state phase transformation from the austenite phase to twinned martensite upon quenching, then to detwinned martensite by mechanical loading, then returning to the austenite phase upon heating. Thus, with the addition of shape memory, these alloys are termed *high-temperature shape memory alloys* (HTSMAs).

To use these alloys in high temperatures, they need to be enhanced in features such as their transformation temperatures, resistance to high-temperature oxidation, material stability at such high temperatures (resistance to precipitation, spinoidal decomposition, and similar phase separations), and low critical stress for slip deformation at elevated temperatures compared with the stress required for martensitic transformation.

Studies have shown that the general operating temperature of SMAs (Ni-Ti) is below 100°C (making them compatible for biomedical use), which is a result of the lower phase transformation temperature and stability in terms of decreased fracture strain and shape memory properties (Luo et al., 2010;

Ortin and Delacy, 2002). Applications such as gas turbines, automobiles, rocket and aircraft engines, and thermal power plants require HTSMMs.

Based on operating temperature, HTSMAs are broadly classified into three categories, as listed in Table 8.1. As observed from the table, the addition of Au, Hf, Pd, Pt, and Zr to Ni-Ti alloy enhances them to HTSMA. Thus, the alloying of SMAs (with transformation temperatures lower than 100°C) with refractory/noble elements (as ternary material—e.g., Ni-Ti-X) allows them to withstand high temperatures and sustain their shape memory behavior.

It can be seen from the table that Ni-Ti alloys have the best performance among other combinations as basic metals in SMAs, although the addition of a third element affects the properties. This explains why most HTSMA studies concentrate around Ni-Ti-based SMAs with the addition of different periodic table elements/metals as ternary components.

The average martensite finish (Mf) temperature of Ti-Ni-Zr and Ti-Ni-Hf HTSMA is around 125°C and the average austenite finish (Af) temperature is around 260°C. This has attracted the interest of material scientists due to the low cost of the raw materials and the relatively high transformation temperatures compared with binary Ti-Ni alloys, thus qualifying them as HTSMAs for specific high-temperature aerospace applications. Apart from high transformation temperatures, the Ti-Ni-Hf-Nb, Ti-Ta-Sn, Ti-Ta-Al, and Ti-Ni-Pd-Cu alloys have good thermal stability and cold workability

TABLE 8.1

HTSMA, Operating Temperature, and Shape Memory Recovery Efficiency

Operating Temperature Range	Alloy Composition	Transformation Temperature (°C)	Recovery (%)
Up to 400°C	Ti-Nb	100–200	97–100
	U-Nb	100–200	–
	Ni-Ti-Hf	100–400	100
	Ni-Ti-Zr	100–250	100
	Cu-Al-Ni	100–400	80–90
	Cu-Al-Nb	100–400	90
	Co-Al	100–400	90
	Ni-Al	100–300	–
	Ni-Mn-Ga	100–400	70
Up to 700°C	Zr-Cu	100–600	44
	Ni-Mn	100–670	90
	Ti-Ni-Pd	100–530	90–100
	Ti-Pd	100–510	88
	Ti-Au	100–630	100
Above 700°C	Ti-Ni-Pt	110–1100	100
	Ti-Pt-Ir	990–1184	40
	Ta-Ru	900–1150	50
	Nb-Ru	425–900	88

compared with binary Ti-Ni shape memory alloys. The Af temperatures of these alloys are below 600°C, which limits the potential for material applications below this temperature. The addition of Pt and Pd has been studied (Yamabe-Mitarai et al., 2015) and was found to be effective in increasing the transformation temperatures of Ti-Ni shape memory alloys (Table 8.1).

As with Ti-Ni binary alloys, Ti-Au-, Ti-Pt-, and Ti-Pd-based alloys also exhibit HTSMA properties due to their enhanced phase transformation temperatures. The shape recovery observed in Ti-Pt is approximately 11% due to the low critical stress for slip deformation compared with the stress required for martensitic transformation. Equiatomic Ti-Pt-based SMAs have an Af temperature around 1000°C, which makes them suitable for operating at high temperatures. The possibility of enhancing their shape memory properties further by increasing the critical stress for slip deformation at high temperatures makes them good candidates for much higher-temperature shape memory material applications.

Note that the critical stress for slip deformation and the critical stress for martensitic transformation/martensite reorientation in SMAs are significantly related to SME. The critical stress for slip deformation should be higher than the stress required for martensitic transformation/martensite reorientation in a good SMA; otherwise, more irreversible deformation will be induced by the dislocation slip, and the shape recovery will be adversely affected. As the temperature increases, the critical stress for slip deformation decreases in SMAs, and once the value is lower than the stress required for martensitic transformation/martensite reorientation, the shape memory properties will drastically decrease, resulting in plastic deformation (efficiency of shape recovery reduces). Research works concentrating on maintaining the critical stress for slip deformation at elevated temperatures are in progress to improve the shape memory properties of Ti-Pt SMAs to qualify as efficient HTSMAs. Thus, Ti-Pt alloys with partial substitutions of Ti with Zr or Ru combinations have been studied, and it was observed that shape memory was enhanced at elevated temperatures in the Ti-50Pt-5Zr and Ti-45Pt-5Ru alloys compared with pure Ti-Pt alloys (Wadood et al., 2012).

Note that during material selection for specific applications, transformation hysteresis plays an important role for SMAs. For repeated active actuation purposes, small hysteresis is desirable, as this will favor the achievable operational frequency and power consumption for actuation. Temperature hysteresis plays a major role in shape memory in high-temperature applications. Large thermal hysteresis during martensitic transformation in SMAs negatively affects the functional stability of SMAs. Hence, SMAs with smaller temperature hysteresis are suitable for actuation purposes, which enables faster recovery. A comparison of the hysteresis of various SMAs is depicted in Figure 8.3. Ti-Ni alloys with ternary elements limit hysteresis to less than 80 K, which makes them potential candidates for high-temperature actuation processes.

Alloying Ti-Pd with Cr has shown improvements in SME with reduced hysteresis at higher transformation temperatures. As discussed earlier for HTSMAs, critical stress for slip deformation at elevated temperatures is favored

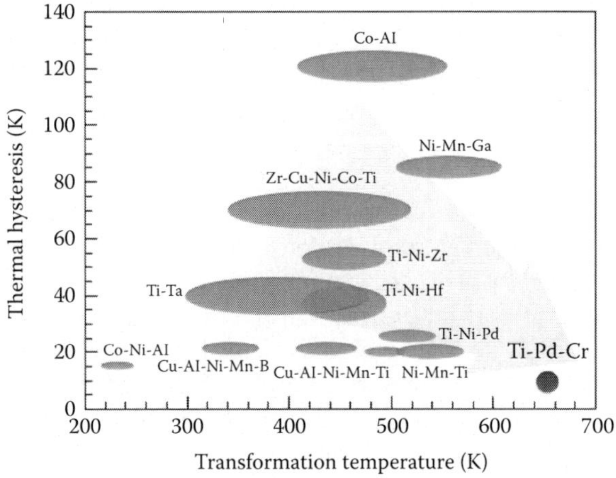

FIGURE 8.3
Plot of thermal hysteresis and transformation temperature for various SMAs. Ti-Pd-Cr alloy shows the highest transformation temperature and lowest hysteresis temperature. (From Xue, D. et al., 2016, Design of high temperature Ti-Pd-Cr shape memory alloys with small thermal hysteresis, *Scientific Reports*, 6: 28244.)

by Ti-Pt alloys and minimal hysteresis is favored by Ti-Pd alloys, and this has narrowed the attention of material scientists to Ti-Pt and Ti-Pd combinations with ternary components from transition metals in the periodic table. Thus, the effect of alloying elements such as Ru, Co, Ir, and Zr on phase transformation and shape recovery has been investigated for Ti-Pd and Ti-Pt alloys (Xue et al., 2016; Wadood et al., 2012). The of addition Ir increased the phase transformation temperature of Ti-Pt and kept the phase transformation temperature of Ti-Pd above 630°C. Another understanding from ternary additions to Ti-Pd and Ti-Pt concerns the improvement of shape memory, which is favorable in the case of adding Zr compared with other transition metals. HTSMAs with Ni-Ti-Pd/Pt alloys have applications in inlet flow control actuations in supersonic passenger flights and flow regulators in spacecraft propulsion systems.

Parallel to SMAs, the growth of polymer science specific to shape memory has opened new horizons in the search for alternatives in high-temperature shape memory in SMPs. Due to the ease of synthesis of SMPs compared with SMAs and the possibilities of inducing new properties by varying stimuli-responsive monomers, researchers have started to look at HTSMPs, although HTSMA studies with ternary metals alloying to Ti-Pd/Pt are still in progress.

8.2 High-Temperature Shape Memory Polymers

Similar to HTSMAs, polymeric monomers are chosen for HTSMPs so that they can survive high-temperature conditions (e.g., thermosetting polymides, polyether ether ketone [PEEK], carbon-enhanced polylactic acid [PLA]).

Polyimides have high thermal stability, excellent mechanical properties, and low dielectric constants, and they are good candidates for various fields, including aerospace, automobiles, microelectronics, and optoelectronics. When compounded with carbon allotropes, polyimide forms composites with high flexural strength up to 350 MPa and flexural moduli up to 21,000 MPa. During high-temperature exposure (of the order 500°C–700°C), the properties are maintained for a short duration, which with improvement would qualify them as high-temperature-grade materials.

High-temperature SMPs have a wide range of applications, such as deployable space structures, shape-morphing structures, smart jet propulsion systems, and high-temperature sensors and actuators. The T_g of the SMP should be higher than the ambient temperature to ensure that the shape recovery process is not triggered prematurely by the operating temperature.

PEEK is another organic thermoplastic polymer, which has an elastic modulus of 3.5 GPa and a density of 1.3 g/cc. PEEK melts at around 350°C and has a glass transition temperature of 150°C. Generally, PEEK is not an SMP, but shape memory behavior in PEEK can be mechanically activated. The T_g and melting points of carbon composites/sulfonated PEEK can be improved, which makes them candidates for high-temperature applications as SMPs.

Figure 8.4 shows the shape memory cycle of a thermoplastic with a high-switching-temperature SMP based on sulfonated poly(ether ether ketone) (SPEEK) ionomers with T_g as high as 200°C for aerospace applications. Three consecutive cycles for Zn-SPEEK/NaOl(30) are indicated here. The T_g of Zn-SPEEK is 208°C and the T_m of NaOl is 250°C, and the difference accommodates a second temporary shape, leading to triple-shape memory behavior. Upon heating to a temperature of 280°C, 57% strain is applied (ε_A), and

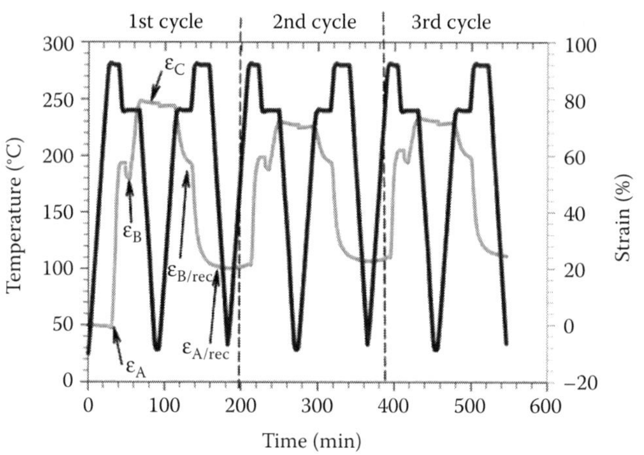

FIGURE 8.4
SPEEK shape memory cycle. (Reprinted from Shi, Y. et al., 2013, High temperature shape memory polymers, *Macromolecules*, 46(10): 4160–4167, with permission from the American Chemical Society.)

upon cooling to a temperature of 240°C, the external stress is removed (corresponding to ε_B). Strain contraction of around 6% can be observed at ε_B. At 240°C, a second tensile deformation of 79% is applied (corresponding to ε_C). Upon applying transition temperature, the two temporary shapes can be utilized for actuation mechanisms in high-temperature conditions.

Similar to HTSMAs, the phase transformation temperature plays a major role in qualifying the SMP as an HTSMP. As discussed in Chapters 5 and 7, T_g is decided by the cross-linking and the monomeric units of the polymer. In the case of an SMPC (analogous to alloying), the properties of individual components in the composites influence T_g and stability at higher-temperature operations. Thermosets, as discussed Chapter 2, have chemical cross-links with better T_g, better moduli across T_g, and greater stability in high-temperature scenarios. Due to the severe working conditions, materials used as aerospace or structural components often require higher modulus and switching temperatures for shape changes and actuation.

Similar to alloying, the addition of carbon allotropes such as graphite, nanotubes, and graphene enhances the properties of SMPs, as shown in previous discussions. Thus, MWNT/poly(biphenyl dianhydride-para-phenylene diamine) (BPDA-PDA) polyimide composite (which has better elastic properties as a composite than as a pure polymer, as discussed earlier) has high thermal stability and a high transition temperature. Thermogravimetric analysis verifies the high-temperature resistance properties of the composite by heating it in a controlled environment. The composite remains stable up to 400°C and only 18% weight loss occurs up to 900°C, indicating the excellent thermal steadiness of the MWNT/BPDA-PDA polyimide composite at high operating temperatures. Studies regarding the strength enhancement of the composite reveal a higher modulus upon enhancement with MWNT, thus proving MWNT/BPDA-PDA as an HTSMP. Thus, the selection of monomers (BPDA and PDA) and their proportions in the polymer composite (MWNT as component of the fiber-reinforced composite) help to improve the transition temperature, which regulates the operating temperature. Apart from the aforementioned properties, another important aspect of HTSMPs is their fatigue strength at high operating temperatures. An HTSMP should maintain its strength and recovery ratios at higher temperatures over continuous cycles of operations. Figure 8.5 compares the storage modulus of a high-cycle-life shape memory polyimide after different bending deformation cycles.

Maleimide-based SMPs qualify as HTSMPs and are well suited to molds for the fabrication of composite structures. Elevated temperatures can be reached inside the mold to cure the resin part without causing dimensional instability in the mold itself.

Thermosetting shape memory cyanate polymers (SMcPs) modified with polybutadiene/acrylonitrile (PBAN) have been synthesized and compared with polyethylene glycol (PEG)-modified SMcPs. These have been studied as strong candidates for use as high-temperature SMPCs with controllable T_g

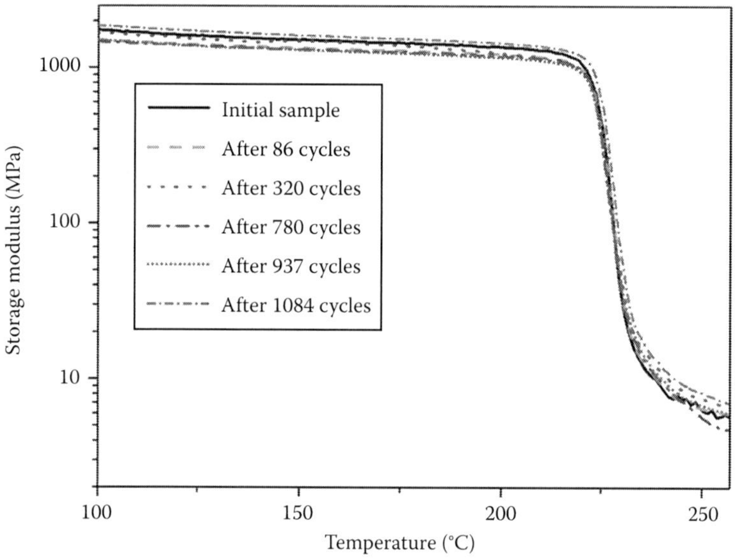

FIGURE 8.5
Plot continuously monitoring the variation in the storage modulus of a shape memory polyimide for 1084 cycles. The storage modulus maintains the trend of crawling along the initial line, and there is much less variation. (From Xiao, X. and Kong, D., 2016, High cycle-life shape memory polymer at high temperature, *Scientific Reports*, 6: 33610: 1–10.)

(Xie et al., 2014). The materials were characterized in terms of their microstructures and their thermal, mechanical, and shape memory properties using various methods, as discussed in the case of HTSMA. The PBAN-modified SMcP had better shape memory properties and higher thermal stability compared with the PEG-modified SMcP. Both the SMcPs showed exceedingly high T_g (above 241.3°C) and higher toughness.

Modified MWNTs in varying weight fractions with good dispersion in the matrix influence the T_g value (details of which will be discussed in the following chapter). Experiments using a poly vinyl alcohol matrix with MWNTs have shown that T_g decreases as the MWNT weight percentage increases beyond a particular content (Du et al., 2015). Thus, having optimal filler content is crucial to improving the T_g value of an HTSMP composite. Cross-linking agents and nanofillers help to synthesize thermoset HTSMPs for space applications. As discussed, the inclusion of nanomaterials not only improves the mechanical properties and thermal conductivity but also enhances electrical conductivity for actuations in remote environments, such as outer space.

Another novel application of HTSMPs is the development of transparent SMPs for lenses and focusing devices in high-temperature applications. Deformable and adaptive optical structures made of transparent shape memory poly(ethylene-co-vinyl acetate) (PEVA) with high thermal resistance/

stability are under scientific investigation due to the demand for reliability and integration into optoelectronic devices (Xiao et al., 2016). Strong intra-/intermolecular charge transfer complex (CTC) interactions, which general aromatic polyimides lack, make them less transparent. Suppressing CTC interactions by incorporating asymmetrical or bulky pendant units, introducing a strong electron-withdrawing fluorinated or sulfone group, and adopting alicyclic moieties in polymer structures can help to make transparent SMP polyimides for high-temperature applications (e.g., lenses/mirrors in spacecraft/scientific space platforms).

Because the outer space temperature fluctuates between –235°C and 150°C in a harsh environment that includes the presence of atomic oxygen, the possibility of sublimation and evaporation at high vacuum, the possibility of cold welding in a frictionless atmosphere, and the presence of high velocity particles, aerospace-grade polymeric materials must be designed considering these factors to account for better performance. Studies regarding the exposure of space-grade SMPs to such severe thermal cycles have been conducted by material scientists across the world, who have reported that HTSMP polyimide had little effect over 200 h due to the harsh conditions in space (Xiao et al., 2015).

Activation by suitable stimuli is also significant based on the application, as discussed in the case of SMPs. Sensitive remote thermal activation may not be possible for HTSMP/HTSMA mechanisms that work in a confined system (electronics) or high-temperature ambiance (space systems); hence, alternate activation mechanisms need to be checked. Electrical/magnetic stimuli activation is a solution for situations where sensitive, precise activation is necessary. Considering the advantages of fast and precise activation without affecting the surrounding systems, electrical activation stands out for SMP actuation.

The following chapter discusses the electroactive mechanisms of SMPCs in detail.

References

Arnold, S. M., 2012. *Materials selection for aerospace: Introduction to aerospace materials; NASA/TM-2012-217411, E-18042*, NASA Glenn Research Center, Cleveland, OH.

Du, F., Ye, E. -Z., Yang, W., Shen, T. -H., Tang, C. -Y., Xie, X. -L., Zhou, X. -P., and Law, W. -C., 2015. Electroactive shape memory polymer based on optimised multiwalled carbon nanotubes/polyvinyl alcohol nanocomposites, *Composites: Part B*, 68: 170–175.

Luo, H., Liao, Y., Abel, E., Wang, Z., and Liu, X., 2010. Hysteresis behaviours and modeling of SMA actuators. In Cismasiu, C. (Ed.), *Shape Memory Alloys*, InTech, Rijeka, Croatia, pp. 61–80.

Ortín, J. and Delaey, L., 2002. Hysteresis in shape-memory alloys, *International Journal of Non-Linear Mechanics*, 37–8: 1275–1281.

Shi, Y., Yoonessi, M., and Weiss, R. A., 2013. High temperature shape memory polymers, *Macromolecules*, 46(10): 4160–4167.

Wadood, A., Takahashi, M., Takahashi, S., and Hosoda, H., 2012. High-temperature mechanical and shape memory properties of TiPt-Zr and TiPt-Ru alloys, *Materials Science & Engineering A*, 2012: MSA29380.

Xiao, X. and Kong, D., 2016. High cycle-life shape memory polymer at high temperature, *Scientific Reports*, 6: 33610: 1–10.

Xue, D., Yuan, R., Zhou, Y., Xue, D., Lookman, T., Zhang, G., Ding, X., and Sun, J., 2016. Design of high temperature Ti-Pd-Cr shape memory alloys with small thermal hysteresis, *Scientific Reports*, 6: 28244.

Xie, F., Huang, L., Liu, Y., and Leng, J., 2014. Synthesis and characterisation of high temperature cyanate-based shape memory polymers with functional polybutadiene/acrylonitrile, *Polymer*, 55(23): 5873–5879.

Xiao, X., Qiu, X., Kong, D., Zhang, W., Liu, Y., and Leng, J., 2016. Optically transparent high temperature shape memory polymers, *Soft Matter*, 12(11): 2894–2900.

Yamabe-Mitarai, Y., Arockiakumar, R., Wadood, A., Suresh, K. S., and Kitashima, T., 2015. Ti (Pt, Pd, Au) based high temperature shape memory alloys, *Materials Today: Proceedings*, 2S: 517–522.

Bibliography

Arockiakumar, R., Takahashi, M., Takahashi, S., and Yamabe-Mitarai, Y., 2013. Microstructure, mechanical and shape memory properties of Ti-55Pd-5x (x = Zr, Hf, V, Nb) alloys, *Materials Science & Engineering A*, 585: 86–93.

Arockiakumar, R., Takahashi, M., Takahashi, S., and Yamabe-mitarai, Y., 2014. X-ray diffraction studies on Ti-Pd shape memory alloys, *Materials Science Forum*, 783–786: 2517–2522

Biju, R. and Reghunadhan Nair, C. P., 2013. Synthesis and characterization of shape memory epoxy-anhydride system, *Journal of Polymer Research*, 20(2): 82.

Hiroshi, A., Koji, N., and Itaru, A., 1979. The glass transition temperatures of copolymers of N-alkyl substituted maleimide and methylacrylate, *Memoirs of the Faculty of Engineering Fukui University*, 27(1) : 49–62.

Kong, D. and Xiao, X., 2016. High cycle-life shape memory polymer at high temperature, *Scientific Reports*, 92: 1–10.

Lan, X., Zhang, R., Liu, Y., and Leng, J., 2011. Fiber reinforced shape-memory polymer composite and its application in deployable hinge in space, *Proceedings of 52nd AIAA/ASME/ASCE/AHS/ASC Structures, Structural Dynamics and Materials Conference*, Denver, CO, pp. 1–10.

Leng, J. S., Lan, X., Liu, Y. J., Du, S. Y., Huang, W. M., Liu, N., Phee, S. J., and Yuan, Q., 2008. Electrical conductivity of thermoresponsive shape-memory polymer with embedded micron sized Ni powder chains, *Applied Physics Letters*, 92(1): 9–11.

Leng, J., Lan, X., Liu, Y., and Du, S., 2011. Shape-memory polymers and their composites: Stimulus methods and applications, *Progress in Materials Science*, 56(7): 1077–1135

Leng, J. and Ye, L., 2009. Smart materials and nanocomposites: Bring composites to the future, *Composites Science and Technology*, 69(13): 2033.

Liu, Y., Wang, X., Lan, X., Lv, H., and Leng, J., 2008. Shape memory polymer composite and its application in deployable hinge for space structure, *Proceedings of SPIE: Sensors and Smart Structures Technologies for Civil, Mechanical, and Aerospace Systems*, 6932(2): 10–18.

Merline, J. D., Reghunadhan Nair, C. P., and Ninan, K. N. 2008. Synthesis, characterization, curing and shape memory properties of epoxy-polyether system, *Journal of Macromolecular Science, Part A: Pure and Applied Chemistry*, 45(4): 312–322.

Mohd Jani, J., Leary, M., Subic, A., and Gibson, M. A., 2014. A review of shape memory alloy research, applications and opportunities, *Materials and Design*, 56: 1078–1113.

Mohr, R., Kratz, K., Weigel, T., Lucka-Gabor, M., Moneke, M., and Lendlein, A., 2006. Initiation of shape-memory effect by inductive heating of magnetic nanoparticles in thermoplastic polymers, *Proceedings of the National Academy of Sciences of the United States of America*, 103(10): 3540–3545

Quackenbush, T. R., Carpenter, B. F., Boschitsch, A. H., and Danilov, P. V., 2008. Development and test of an HTSMA supersonic inlet ramp actuator, *Proceedings of SPIE: The International Society for Optical Engineering*, 69300: 11 pp.

Ragin Ramdas, M., Santhosh Kumar, K. S., and Reghunadhan Nair, C. P., 2015. Synthesis, structure and tunable shape memory properties of polytriazoles: Dual-trigger temperature and repeatable shape recovery, *Journal of Materials Chemistry A*, 3(21): 11596–11606.

Santhosh Kumar, K. S., Biju, R., and Reghunadhan Nair, C. P., 2013. Progress in shape memory epoxy resins, *Reactive and Functional Polymers*, 73(2): 421–430.

Sokolowski, W. M., 2004. US Patent no. US 006702976 B2: Cold hibernated elastic memory self deployable and rigidizable structure and method therefor, March 9.

Wadood, A., Arockiakumar, R., and Hara, T., 2014. High-temperature shape memory alloys based on Ti-platinum group metals compounds, *Materials Science Forum*, 783–786: 2541–2545.

Wu, X., Huang, W. M., Zhao, Y., Ding, Z., Tang, C., and Zhang, J., 2013. Mechanisms of the shape memory effect in polymeric materials, *Polymers*, 5(4): 1169–1202.

Xiao, X., Kong, D., Qiu, X., Zhang, W., Liu, Y., Zhang, S., Zhang, F., Hu, Y., and Leng, J., 2015. Shape memory polymers with high and low temperature resistant properties, *Scientific Reports*, 5(2): 14137.

9

Electroactive Shape Memory Polymer Composites

The current focus of research into shape memory polymer composites (SMPCs) is oriented toward improving recovery stress (depending on the modulus), recovery speed (primarily due to poor thermal conductivities), and inertness to electromagnetic stimuli (due to the electrical insulation of most polymeric materials). To overcome these limitations, fillers are added to improve the rubbery moduli and thermal and electrical conductivities.

Thermally activated actuators can offer large strains at the expense of large amounts of energy and slow response. Magnetically driven shape memory devices tend to be large in size due to the intricate mechanisms involved, while electrostrictive (electric field–driven) shape memory devices are more efficient, have faster response times, are compact in size, and solve most of the problems experienced by other stimuli-responsive shape memory materials.

9.1 Electrically Stimulated Shape Memory Materials

Conventional thermoresponsive SMPCs use heat energy to achieve temperatures above T_g for phase transformation, and hence a large amount of energy is used to raise the temperature of the bulk. In the case of electroactive triggering, the conductive part alone is heated by resistance, and hence the Joule heating of *electroactive shape memory polymers* (EASMPs) proves to be more efficient than that of thermoresponsive SMPs. When a polymer composite exhibits shape memory properties upon the application of an electric current/field, it is called an *electroactive shape memory polymer composite* (EASMPC). The electroactive shape memory effect (SME) typically depends on the amount of induced Joule heating, and it can only occur in a polymer when the temperature is above T_g. Induced Joule heating is proportional to the applied voltage and inversely proportional to the resistance of the materials. A faster electroactive response is associated with an improvement of the thermal and electrical conductivity of the composites. The electrical conductivity of such polymers depends on the dispersion of fillers in the continuous phase during polymerization and the monomers involved in the process. Thermal conductivity and T_g show strong dependence on the weight fraction of filler materials in SMPs. The inclusion of conductive polymers or filler materials can improve the electrical/thermal conductivity of the bulk and thereby enhance electrical actuation.

A polymer can be synthesized to be intrinsically electrically conductive either by using monomers with good electrical properties or by doping conductive fillers (carbon nanotubes [CNTs], carbon nanofibers [CNFs], graphite, graphene, and metallic particles) in the insulating polymer and finally initiating SME by Joule heating.

9.1.1 Intrinsically Conductive Polymers

Organic polymers that conduct electricity are called *conductive polymers*, and since the property is inbuilt, they are *intrinsically conductive polymers*. In order to show electrical conductivity, a polymer has to mimic a natural conductor (metal) by allowing electrons to move freely inside it. This is chemically possible in an organic material due to the conjugate nature of the bond between carbon atoms, as the potential polarization results in electron movement. In conjugation, the bonds between the carbon atoms are alternately single and double. Every carbon–carbon linkage contains a localized *sigma* (σ) bond or a strong covalent bond and a weaker localized *pi* (π) bond or double bond. This leads to the shifting of electrons by breaking weaker bonds, resulting in the release of electrons in the presence of a suitable medium. One of the simplest forms of conjugated polymer is polyacetylene, which is formed by the polymerization of acetylene.

Figure 9.1 depicts the chemical structure of a few (conjugate bond) polymers that are intrinsically electrically conductive: polyacetylene, polyphenylene vinylene, polypyrole (X = NH), polythiophene (X = S), polyaniline (X = NH/N), and polyphenylene sulfide (X = S).

The component monomers having conjugated chemical bonds (as in Figure 9.1) under certain conditions (doping leading to protonation) ensures the electron conductivity of the polymer. For example, when polyaniline (PANI) in an emeraldine oxidation state is treated with acids, the protons primarily interact with the imine atoms of nitrogen (protonation), and this results in formation of polycations. This causes the localization of positive charges on adjacent nitrogen atoms (see the polyaniline structure in Figure 9.1), increasing the total energy of the system. This positive charge accumulation increases the electron density, causing the unpairing of lone electron pairs of nitrogen atoms, thus increasing the material's conductivity.

9.1.2 Fillers for Electroactivity

Conductive filler materials in the form of fibers or powder enhance the heat transfer in the bulk matrix of the polymer and thus facilitate electrical conduction. Synthesis of conductive SMPC by mixing various kinds of conductive fillers such as short carbon fibers (SCFs), carbon black, metallic (Ni powder) powders, and CNT is being studied by material scientists across

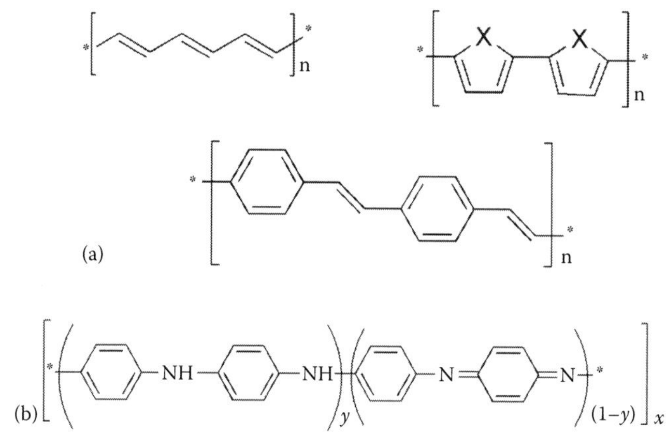

FIGURE 9.1
(a) Chemical structures of a few intrinsically conductive polymers. Here, the conjugate bond system provides suitable polarity-releasing electrons by breaking weaker bonds. (b) Structure of polyaniline, a widely used conducting polymer. (The redox state of the polymer is determined by the value of y, which may vary continuously from zero to unity. At $y = 0.5$, polyaniline occurs in the form of emeraldine; $y = 0$ corresponds to the fully oxidized form, pernigraniline, while $y = 1$ corresponds to the fully reduced form, leucoemeraldine.)

the world (Lu et al., 2013; Aguilar et al., 2010; Lan et al., 2008; Park et al., 2008; Zhang et al., 2009). Initiation of the same has been studied by means of remotely controlled Joule heating, which provides the initial insights on the electroactivity of SMP with reinforcement fillers.

Nickel powder in various polymer matrices has been tested as metal–polymer composite shape memory material, and optimal Ni dispersion has been tested for various matrix materials. Studies report variations in the T_g of epoxies due to the presence of filler materials (Santhosh Kumar et al., 2013). As the content of filler goes beyond the critical requirement, it becomes the dominant phase and regulates the properties of the composite. Hence, the weight percentage of filler material is significant in the case of a doped conductive SMPC. Apart from inducing conductivity, carbon in various forms as reinforcement filler in the SMPC will positively influence various properties, such as the modulus and T_g.

Carbon is unique, with allotropes such as diamond, amorphous carbon, fullerenes, graphite, and CNTs/CNFs all having good mechanical, thermal, and electrical properties. This justifies the use of carbon-based nanostructures in SMP nanocomposites for their electrical sensitivity. The following sections discuss the usage of CNFs, carbon nanoparticles, CNTs, and carbon nanopaper (CNP) as fillers in polymer matrices to transform them into EASMPs.

9.1.2.1 Carbon Nanofibers

CNFs are SCFs known for their potential enhancements to the thermal, electrical, frequency shielding, and mechanical properties of composites. Thereby, filler material influences not only the electrical properties but also the mechanical and thermal properties of a polymer. CNFs have better properties than other carbon fibers in use due to their size advantage, which provides a larger surface area, a continuous internal network of fibers, and greater strength and modulus. Shape recovery for EASMPs with CNFs can be accomplished by applying an external electric current through leads connected to the material.

SMP nanocomposites with excellent high-speed electrical actuation capabilities have been developed by many material scientists by incorporating electrospun continuous nonwoven CNFs into an epoxy-based SMP matrix. Nonwoven nanofibers commonly have large surface areas and high porosity. Electrospinning is one of the most effectual techniques for synthesizing nonwoven continuous polymer nanofibers of diameters ranging from several micrometers down to a few nanometers. These fibers have properties similar to fibers produced by electrospraying and conventional solution dry spinning. An electrically percolating network is provided by the morphology of the nanosized fibers, which results in high electrical conductivity, enhanced heat transfer, and better recovery stress.

Studies with epoxy shape memory polymer (ESMP) resin systems as matrices and CNFs as reinforcement have shown reduced SME with increased CNF content. This is because of high interfacial friction between the ESMP macromolecular segment (the matrix) and the fillers. This emphasizes the need to optimize the nanofiber content for specific properties. Thus, it can be inferred that filler content is a trade-off between the interfacial bond strength of the matrix and the reinforcement and the electrical properties. Note that, unlike Ni powder, the incorporation of CNFs has little effect on transition temperature. The ESMP–CNF composite shows that the recovery speed is influenced by the narrow glass transition of the epoxy, and the increased thermal conductivity is due to the incorporation of CNFs.

As discussed in Chapter 5, the range of transition is a function of the molecular mobility of polymer chains, which in turn is a function of the filler content. Optimal filler content reduces resistance to molecular mobility due to the reduced viscosity of the matrix, and thus the transition happens over a shorter range of temperatures. A narrow glass transition helps in efficient recovery and stable phases across the T_g.

Increasing the triggering voltage also accelerates the shape recovery (Figure 9.2); as demonstrated through an experiment on an ESMP–CNF composite, a recovery time of less than 2 s is achievable by applying an increased triggering voltage (Luo and Mather, 2010). A comparison of the recovery percentages for various applied voltages can be seen in Figure 9.2a, which clearly indicates that increasing the voltage lowers the recovery time. A plot of characteristic induction times and recovery times against the applied voltages is shown in Figure 9.2b, which shows a direct relation between time and

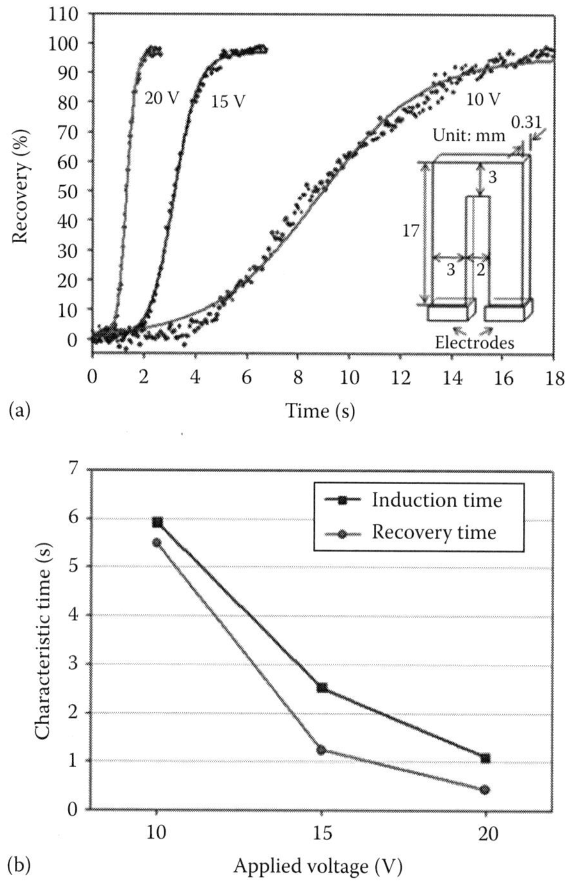

FIGURE 9.2
Electrically activated shape recovery of EP–CNF nanocomposites made in the shape of Π. (a) Time–shape recovery plot at various voltages; a shorter recovery time trend is obtained as the stimuli voltage increases. (b) Induction–recovery time plot against stimuli voltage and recovery voltage; the profile takes less characteristic time against the applied voltage. (Reprinted from Luo, X. and Mather, P. T., 2010, Conductive shape memory nanocomposites for high speed electrical actuation, *Soft Matter*, 6(10): 2146–2149, with permission from the Royal Society of Chemistry.)

voltage. *Induction time* is the time it takes to initiate a reaction. In the case of an electroactive material, it is the time required for the Joule heating to be initiated (i.e., for the material to be heated up), and this is a function of the thermal conductivity of the material.

When plotting induction and recovery times for the three DC voltages, for a given voltage, the induction time is always longer than the recovery time. Induction and recovery times are reduced with the applied voltage.

9.1.2.2 Carbon Black

CB is a commercial form of solid carbon in the size range 10–500 nm; it typically contains more than 95% pure carbon and is balanced with trace elements such as oxygen, hydrogen, and nitrogen. When CB is added as a reinforcement filler, the resilience, tear strength, conductivity, and other physical properties of the polymer matrix composites are improved.

Combinations of CB and SCFs are used in many experiments to study the electrical properties of SMP composites. The materials that typical receive the most attention are styrene-based thermoset polymers. The literature reports an increase in resistance with an increase in temperature (positive thermal coefficient [PTC]) with the addition of 5% CB, while the addition of a combination of SCFs and CB in styrene-based resin decreases resistance as the temperature increases (negative thermal coefficient [NTC]). Thus, an SMPC with CB–SCF reinforcements acts as a thermistor, which regulates resistance based on filler content. This is significant, as a combination of CB and SCFs could increase volume conductivity and provide better shape memory by enhancing Joule heating by conduction. The positive thermal coefficient effect for conducting polymer composites depends heavily on the properties of polymer matrices and conducting fillers.

Styrene- and CB-based SMPs have their T_g reduced with increases in CB content from 0% to 10% (69°C–56°C). Electrical resistivity is extremely high in CB content less than 2%, but a sharp transition occurs between 2% and 6% (referred to as the *percolation threshold*). Thus, while keeping the CB content at 5%, the other filler content is varied to understand the synergy of reinforcements in a particular matrix.

Another observation about this composite is that the combination of CB and SCFs decreases the T_g value, as the SCF content increases from 0% to 2% with 5% constant CB content. An important inference from studies like Meng and Hu (2009) concerns the inherent fibrillar form of SCFs/CNFs, which have a higher tendency to form a three-dimensional network in the composites, ensuring better electrical responses than with CB as filler. Fibrils are slight thread-like projections from the sides of SCFs, making it an imperfect surface material (it has an *undulated* surface). These undulations and projections are advantageous as links to the adjacent SCF entity (*interfiber linkages*), making a network possible in three dimensions. Hence, individually, as the amount of SCF filler content increases, the electrical conductivity of the SMPC also increases with less recovery time, which is attributed to the fibrillar form, as explained above.

A higher weight percentage of CB in polyurethane SMP shows better strain fixity, but the further addition of CB decreases the shape recovery ratio and shape recovery speed, while T_g remains unaffected. As the CB content increases above 10%, the crystallinity decreases, which affects the ability of the deformed sample to retain its shape after the removal of the load. This reduces the shape fixity ratio and thus the recovery ratio as well.

The effect of CB nanoparticles (with no SCFs) on two-way reversible shape memory cycles in cross-linked polyethylene is improved shape memory behavior. The development of a stable physical cross-linking structure preserves the elastic energy in the recovery process in a CB-reinforced polymer matrix.

Two strategies have been utilized in manufacturing electrically conductive SMP composites with CB, the first being the direct addition of CB into a polymer matrix, while other is to use a hybrid CB filler (CB with any other filler: SCFs/CNFs, Ni powder, etc.) to synergistically lower the electrical resistivity of the resulting SMP.

Figure 9.3a–c shows the SME of a random dispersed fiber composite matrix with three different combinations of CB and Ni particles. The particles are interconnected internally, forming conductive networks that result in the electrical conductivity of the composites.

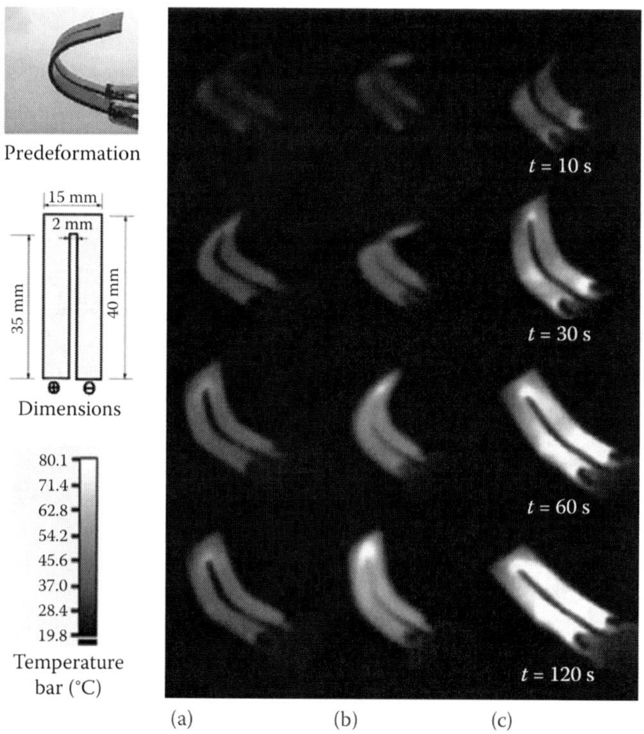

FIGURE 9.3
Shape recovery with temperature distribution via IR observation of three combinations of Ni and CB. (a) 10 vol% of CB alone. (b) 10 vol% of CB, 0.5 vol% of randomly distributed Ni. (c) 10 vol% of CB, 0.5 vol% of chained Ni). The dimensions of the sample and the temperature variations are shown at the side. (Reprinted from Leng, J. et al., 2011, Shape memory polymers and their composites: Stimulus methods and applications, *Progress in Material Science*, 56(7): 1077–1135, with permission from Elsevier.)

It can be inferred that the continuity of well-dispersed reinforcements improves the overall electrical conductivity. The original shape is recovered within 120 s in 10 vol% of CB and 0.5 vol% of chained Ni particles. The performances improve as the network of nickel is arranged/chained to form a continuous network.

The effect of strain on electrical resistivity is significant, as SMPCs are suitable candidates for strain sensors. Similar to induced polarity in barium titanate piezoelectric material, as explained in Chapter 1, the external stress results in a strain in the material that internally connects the networks, resulting in electrical resistivity changes as the strain increases. This makes SMPCs competent alternatives to piezoelectric materials. The urge to improve the properties of composites to supplement smart materials leads to the search for better alternative fillers for EASMP materials. As explained in the following section, this leads the research into areas such as nanotubes.

9.1.2.3 Carbon Nanotubes

Dr. Sumio Iijima was the first to observe tubular carbon structures with diameters of approximately 1 nm and large aspect ratios. Generally, MWNTs and SWNTs are produced by three techniques: arc discharge, laser ablation, and catalytic growth. For practical applications, CNTs are categorized by their elastic response, inelastic behavior/buckling, yield strength, and fracture. Composite materials, which are the major players in aerospace engineering, will be most benefited if the properties of CNTs are combined with the conventional materials in use. CNTs possess an average Young's modulus value of 1.8 TPa, which is much higher than that of conventional carbon fibers (680 GPa) used as reinforcements. MWNTs synthesized by chemical vapor deposition show thermal conductivity of 4–300 K from room temperature to elevated temperature conditions. The specific strength of CNTs is extremely high compared with traditional fibers, and they are comparable to metals as conductors of heat and electricity. CNTs consist of stacks of tens or hundreds of concentric shells of carbon separated by approximately 0.34 nm. This is analogous to the honeycomb arrangement of carbon atoms in graphite sheets, and CNTs have very high Young's modulus and tensile strength, making them preferable for composite materials with improved mechanical properties.

The various aspects of CNTs as either single-walled or multi-walled nanotubes (SWNTs/MWNTs) have been explored for their use as reinforcements in polymer composites to enhance the mechanical/thermal/electrical properties in various applications. As discussed earlier, aligned fibers give better results in electroactivity. Figure 9.4 is a conceptual visualization of CNT reinforcements (the tubular structures) in a polyethylene matrix (the shaded objects are polyethylene chains).

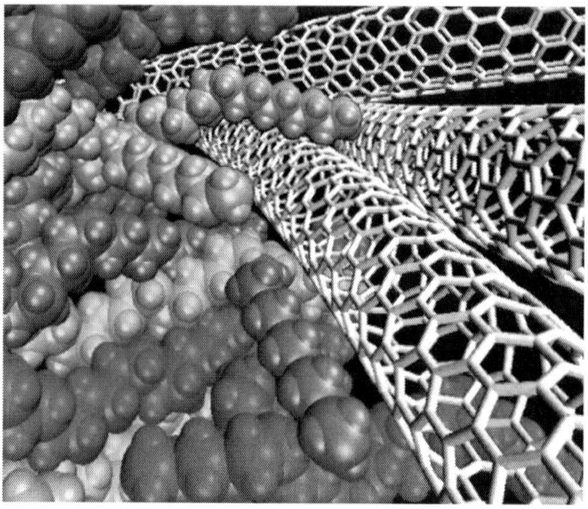

FIGURE 9.4
Conceptual image of SWNTs (right) bundled together when used as the reinforcing element of a composite material with CNT polyethylene chains (left). (Reprinted from Ajayan, P. M. and Tour, J. M., 2007, Nanotube composites, *Nature*, 447(28): 1066–1068, with permission from Macmillan Publishers Ltd.)

Nanotubes are graphite sheets rolled into perfect hollow cylinders that range in the order of 0.5–2.0 and 2–50 nm in diameter for SWNTs and MWNTs, respectively.

Nanotubes increase stiffness, strength, and toughness and enhance electrical and thermal conductivity when it is reinforced in a matrix. If they are efficiently aligned at the macroscopic scale, CNTs can replace continuous carbon fibers, which can be woven and stacked to create extremely strong, high-integrity fabrics for engineering applications. As the dimensions are the smallest of existing fibers, the increased surface area in matrix interactions will favor the properties of composites. Nanotubes about 10 nm in diameter have an interfacial area with matrices almost 1000 times greater than that of 10 μm fibers of the same volume.

A significant concern expressed in the literature (Jiang et al., 2014) regarding the use of CNTs as reinforcements for enhancing electrical properties is the challenge of preserving the highly electronically active surface of CNTs (not covered by a surfactant) and to make sure that the CNTs do not break under shear exceeding a certain threshold, due to their extremely high aspect ratios. Another major drawback of CNTs as reinforcement materials is their pronounced tendency toward agglomeration, making their homogeneous integration (dispersion) into a matrix difficult. Surface modifications are a solution to these issues, and this technique has improved the interfacial properties of the SMP–CNT system, for which many researchers have used nitric acid and sulfuric acid to treat the surfaces of nanotubes. Upon mixing surface-modified CNTs into a polyurethane SMP and applying an electrical voltage, the surface modification increases the matrix–reinforcement bondage and improves the electrical properties.

Figure 9.5 shows the dispersion of surface-modified CNTs in a sulfonated poly(styrene-b-[ethylene/butylene]-b-styrene (SSEBS) matrix, as observed through scanning electron microscopy (SEM). The composite SSEBS–CNT is tested for SME, and it shows better results compared with nonmodified CNTs in the same matrix.

Five strategies have been adapted to induce electroactivity in CNT-reinforced polymer matrices.

1. Mixing CNTs directly into the polymer matrix
2. Blending surface-modified CNTs with the polymer matrix, which favors interfacial bonding between the matrix and the reinforcement
3. Converting CNTs in paper or film form and incorporating them into the polymer matrix
4. Aligning the CNTs in the polymer matrix in the presence of an electromagnetic field
5. Cross-linking CNTs with the polymer matrix

The major challenges in this area include the dispersion of CNTs in the polymer matrix, bonding between the polymer macromolecules and CNTs, electrically conductive networks in composites, and the electrical properties of composites.

Studies have proven that CNTs are one of the more effective fillers for making SMPs electrically conductive. Many research groups have investigated electroactivity, and the SME of any polymeric materials is dependent on the

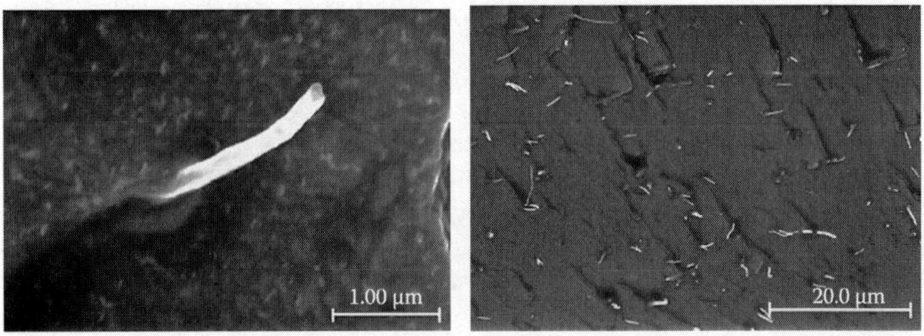

(a) (b)

FIGURE 9.5
SEM image of CNF reinforcement in an SSEBS composite matrix. (a) Separate CNFs in the matrix. (b) Dispersion of CNFs in the matrix. (Reprinted from Wang, X. L. et al., 2009, Enhanced electromechanical performance of carbon nanofiber–reinforced sulfonated poly(styrene-b-[ethylene/butylene]-b-styrene) actuator, *Composites Science and Technology*, 69(13): 2098–2101, with permission from Elsevier.)

filler content, the degree of surface modification of the fillers (in the case of CNTs), the mode of synthesis, and the matrix properties (Aguilar et al., 2010; Lan et al., 2008; Meng and Hu, 2009; Wang et al., 2015; Wang et al., 2009; Lu et al., 2013; Park et al., 2008; Zhang et al., 2009). At the same filler concentration, surface-modified MWCNT composites showed better electrical conductivity than untreated samples.

In situ polymerized structures have shown better (aligned) distribution of fillers/CNTs compared with direct mixing. Electrical conductivity is enhanced 100-fold for chained/aligned CNTs compared with randomly distributed fibers in the polymer matrix. As discussed, aligned CNTs show better (isotropic) properties compared with randomly distributed fibers. This is because the alignment of the nanotubes provides a more effective path for electron transfer along the length of the composite (in the direction of alignment).

Quantification of the electrical properties is required as the material choice becomes significant for space-grade materials. The general method employed for resistivity/conductivity is four-point probe measurement, which eliminates the lead resistance. During resistive heating, an infrared camera can be employed to investigate the temperature distribution of the SMPs during the shape recovery process.

The distribution of fillers plays a major role in forming the percolation network in the matrix, and this can be verified microscopically. Figure 9.6 shows a comparison of microscopic images of specimens with and without CNTs as reinforcement. There are surface-modified as well as nonmodified CNT fillers, and in the surface modification itself two varieties of modification agents were used. Specimen P (Figure 9.6d) represents the glass fiber–epoxy (GF–EP) composite (the non-CNT sample), specimen J2 represents the SWCNT–GF–EP composite (Figure 9.6a), specimen J3 (Figure 9.6b) represents SWCNT–GF–EP composites with the SWCNTs treated with volan (a chemical surface treatment agent), and specimen J4 (Figure 9.6c) represents SWCNT–GF–EP composites with the SWCNTs treated with BYK-9076 (a wetting and dispersing agent).

It can be inferred from the microstructure that in Figure 9.6a, CNT and GF are segregated out in the EP matrix. The composite treated with volan alone has dispersed the CNTs in the matrix better than the untreated sample, but GF agglomeration continues. When using volan and BYK-9076, both the CNT and GF networks remain continuous even though they are well dispersed and properly bonded with the matrix system.

A strength comparison of the three samples (J2, J3, and J4) with the composite without CNTs as components (P) shows a gradual increase in the flexural strength of the composite system. The probable reason for the increase in strength of specimen J4 is the improved bonding and interfacing between the SWCNTs and GF–EP. Thus, surface modification affects dispersion and strength too.

The tensile strength and modulus of SWCNTs have been measured as high as 200 GPa and 1 TPa, respectively, but the exact stresses that are borne by these

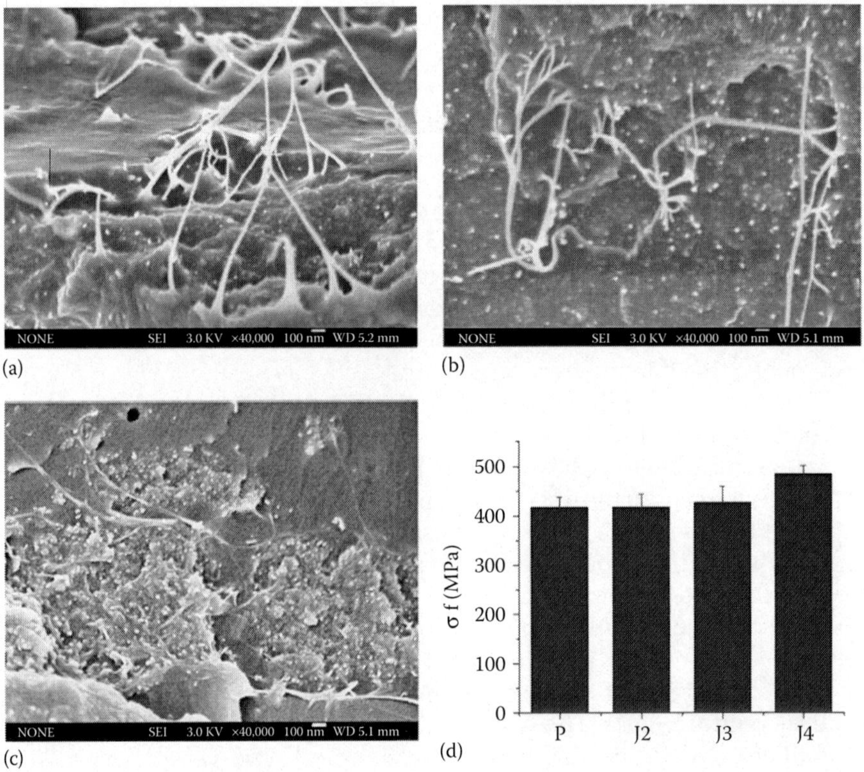

FIGURE 9.6
(a–c) SEM images at ×40,000 magnification for samples J2, J3, and J4, respectively. (d) Variation of flexural strength in each sample with varying dispersion agents. (Reprinted from Yan, Z. et al., 2009, The dispersion of SWCNTs treated by dispersing agents in glass fiber–reinforced polymer composites, *Composites Science and Technology*, 69(13): 2115–2118, with permission from Elsevier.)

high-strength SWCNTs in the EP matrix are unknown because of uncertainty in the understanding of the interface between SWCNTs and the surrounding matrix. A sufficient stress transfer length is also important to allow all the stress to be transferred from the matrix to the SWCNT. Since the modulus of a polymeric material with shape memory and the dynamic mechanical properties of a polymer are described in terms of a complex dynamic modulus (E^*), which is a complex combination ($E^* = E' + i\, E''$) of storage (E') and loss modulus (E''), the effect of dispersion/bonding on the storage/loss modulus for the specimen (P, J2, J3, J4) is clear. The tan δ (E''/E') variations of different dispersion agents for increase in temperature, as depicted in Figure 9.7, are a result of the dynamic mechanical analysis (DMA) plot.

Even though the T_g of the system of composites is not influenced by the addition of SWCNTs and surface treatment methods, the tan δ value increases for P and J2 at the transition temperature, which may be due to an increase in loss modulus (rubbery elastic modulus) at T_g. This signifies that the presence

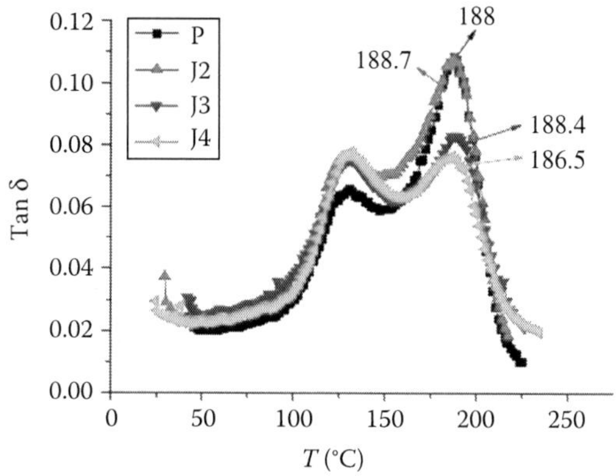

FIGURE 9.7
Tan δ–temperature plot for SWCNT–EP–GF composite system with and without dispersing/surface treating agent. (Reprinted from Yan, Z. et al., 2009, The dispersion of SWCNTs treated by dispersing agents in glass fiber–reinforced polymer composites, *Composites Science and Technology*, 69(13): 2115–2118, with permission from Elsevier.)

of dispersing agents and surface treatments in J3 and J4 provides a stable modulus across T_g.

The electrical properties of nanotubes in composite systems suggest that the electrical conductivity of surface-modified MWNT composite is lower than that of composites filled with untreated MWNTs at the same filler content. This is attributed to increased defects in the lattice structure of the carbon–carbon bonds on nanotube surfaces due to acid treatment/modification. Thus, even though strength and interfacial bonding is favored by surface treatment, the negative effect on electrical behavior reduces the shape memory response of the composites. Hence, it is established that SME depends on the filler content and the degree of surface modification of the MWNTs.

An interesting report regarding the addition of CNTs in polymers concerns the sequence of blending components during the polymerization process (Qi et al., 2016; Zhang et al., 2009). The sequence of blending affects the dispersion of CNTs, as the viscosity changes every moment of polymerization. Changes in the viscosity of the matrix affect the dispersion of reinforcement fibers, which plays a vital role in the electrical properties of the SMP matrix.

The electrical conductivity of the CNT-reinforced composite matrix varies with the filler content, as previously discussed. Other than filler content, monomer content also affects the properties of the final polymeric product. This has been verified by the results obtained in the case of a composite matrix made of poly(propylene carbonate) (PPC) and poly(lactic acid) (PLA) as monomeric units and MWNTs as the reinforcement (Qi et al., 2016).

The electrical conductivity was verified for the PPC–PLA–CNT composites at different MWNT concentrations (with 1, 2, and 3 parts per hundred PPC and PLA by weight concentrations of MWCNTs). The composite matrix is made up of monomeric units of PPC and PLA in ratios of 70:30 and 50:50 by weight percentage.

As discussed earlier, the electrical conductivity increases when the nanoparticle content increases (Figure 9.8a,b). Different triggering voltages are applied for electrical actuation and the variations in the complimentary surface temperatures and recovery times are noted.

It can be inferred that the applied voltages (20 and 30 V) and parts per hundred CNTs (CNT1, CNT2, CNT3) in the matrix are both directly proportional to the recovery times in the case of PPC70–PLA30 composite (Figure 9.8a).

Except for 1 phr CNTs in both cases, other combinations show improvements with matrix optimization. Note that as the CNT content changes from 1 to 3 phr, 30 V stimulation shows a better response in the case of the 70:30 matrix, while in the 50:50 matrix, it reverses except for CNT3. That proves that at lower CNT concentrations, lower voltages can give a better response for an optimized matrix composition of PPC–PLA. When the CNT concentration increases (3 phr and above), higher voltages can give a faster heating response for optimized matrices.

CNTs as networks behave as metallic, semiconducting, polymer-like materials, which makes them the most advantageous smart material for aerospace applications, where the properties of metals, ceramics, and polymers need to be combined. Apart from aerospace and other high-temperature applications, the properties of EASMPs are suited to robotics, self-healing structures, biomedical applications, and more.

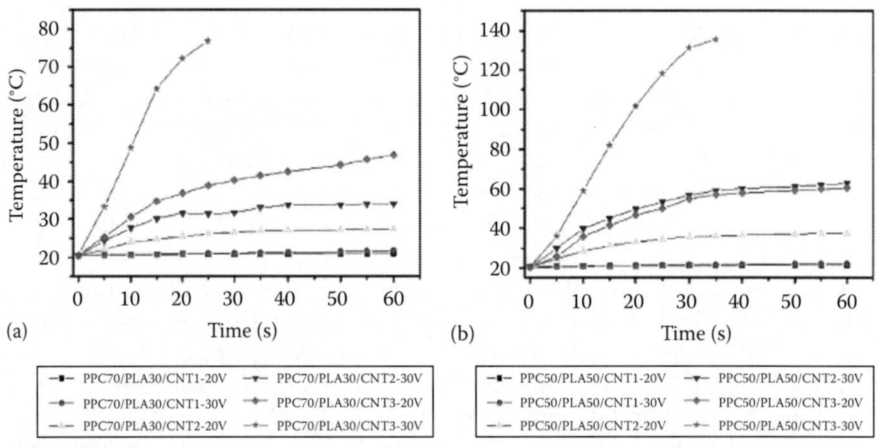

FIGURE 9.8
Triggering voltage (20, 30 V) draws the recovery time–surface temperature relation for two ternary combinations (PPC–PLA–CNTs). CNT1/CNT2/CNT3 denotes 1, 2, 3 phr of CNTs. (Reprinted from Qi, X. et al., 2016, Selective localization of multi-walled carbon nanotubes in bi-component biodegradable polyester blend for rapid electroactive shape memory performance, *Composites Science and Technology*, 125: 38–46, with permission from Elsevier.)

Many bioapplications of electroactive SMPs with tunable T_g near human body temperature have been reported (Deng et al., 2016). Much of the attention is on tissue engineering based on poly(e-caprolactone) (PCL) with different molecular weights and conductive amino-capped aniline trimer. Through these studies, material scientists and biologists together could demonstrate the potential to enhance myogenic differentiation from C2C12 myoblast cells (a mouse cell line typically used for investigating the growth and differentiation of muscle). Myogenesis is the process of forming muscular tissue, particularly during the embryonic development stage. Muscle fibers are formed from the fusion of myoblasts (a mononucleate cell type that fuses with other similar mononucleates), resulting in multinucleated fibers called *myotubes* that eventually develop into skeletal muscle fibers. In the early development of an embryo, myoblasts can either proliferate or differentiate into a myotube.

The potential engineering applications of EASMPs in skeletal and muscle tissue are being explored. The biological applications of EASMPs demand biocompatibility as well as biodegradability for obvious reasons. Thus, studies of EASMPs can be advantageous toward biodegradability as well as biocompatibility. Biodegradability is mainly governed by the monomeric units involved in SMP polymerization. During the formation of a composite material with electroactive SMPs as the matrix, the choice of biodegradable monomers can lead to composite SMPs with biodegradable properties and thus biocompatible polymers with shape memory properties.

9.1.2.4 Carbon Nanopaper

The introduction of carbon-based bulk material in SMPCs to facilitate electroactivity has been a significant milestone, and thus the electric triggering of SMPCs enlarges their technological potential. In the previous section, the use of nanofibers, nanotubes, and nanoparticles to reinforce SMPs with various polymers as matrices in different weight fractions, microstructures, and their synthesis and applications were explored. Another advent in this area is the use of nanopaper, which has been studied and explored in detail by various material science enthusiasts across the world (Wang et al., 2015; Lu et al., 2014).

A planar conductivity was intended while shifting the attention to nanopaper from one-dimensional conduction as imparted by nanoparticles, fibers, and nanotubes. This is thought to improve volume conductivity more than the use of nanoparticles/-fibers/-tubes. Studies on multiple layers of nanopaper incorporated into SMP matrices have shown improvements in electrical properties and faster actuation with less triggering voltage (Lu et al., 2014).

The major approach toward the synthesis of CNP is from *graphene* or single-layer carbon foil, and the other route is by stacking together CNT/CNF films known as *buckypaper*. Vacuum-assisted resin transfer molding is used to synthesize nanopaper from CNT/CNF layers. Buckypaper is one-tenth the weight of steel but around 500 times stronger when its sheets are stacked to form a composite.

A specimen of reduced graphene oxide paper (RGOP) used as reinforcement in an EP resin matrix was electroactivated and observed for its temperature distribution in the bulk by means of infrared thermal imaging (Figure 9.9a). A 6 V DC current was applied for electrical stimulation, which induced Joule heating in the sample and showed faster bulk heating than that in one-dimensional reinforcement filler SMPC. Compared with the issues regarding the dispersion of composites with reinforcements such as nanofibers, particles, and nanotubes, a uniform layer can be obtained in the polymer as a laminar composite in a nanopaper composite system. Figure 9.9a shows the timeline of the heating process, with the temperature range stretching from room temperature to 120°C. Within a short period (5 s), the entire area of the specimen reaches 120°C by Joule heating. Thus, it is superior in heat transfer compared with particle-/fiber-reinforced composite systems. Figure 9.9b is a three-dimensional depiction of the temperature distribution in the bulk over a cross-section. A uniform distribution of temperature is achieved in the bulk, which is an interesting result in terms of the future of thermally/electrically activated shape memory materials.

RGOP, being an excellent conductor of heat, can act as a conductive layer in the matrix and transmit heat. Uniform heat distribution over short periods results in faster recovery. This can be explained as the result of increased system entropy that is proportional to the temperature due to increased triggering voltage. A continuous conductive network formed by self-assembled and multilayered CNF nanopaper shows better Joule heating to an increasing number of layers of nanopaper. Thus, the more CNFs there are in the nanopaper, the more conductive paths there are forming a continuous network for charge transfer, resulting in better shape recovery efficiency.

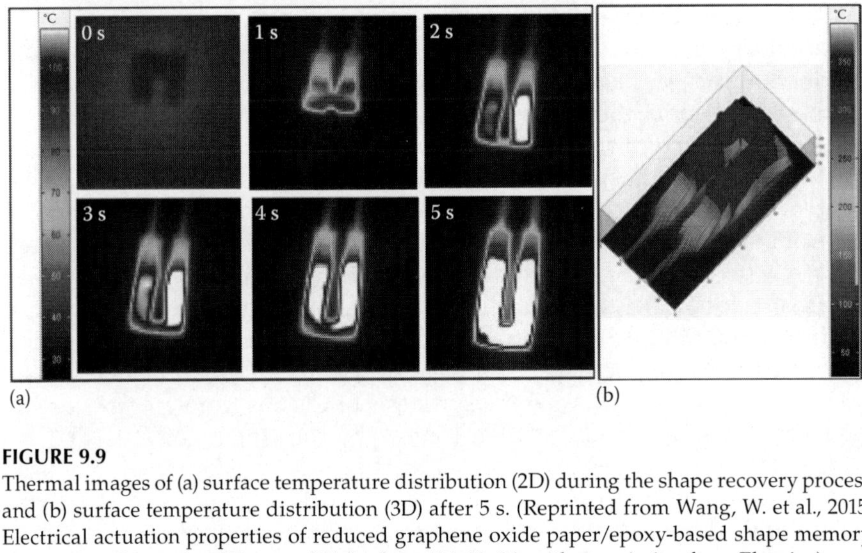

FIGURE 9.9
Thermal images of (a) surface temperature distribution (2D) during the shape recovery process and (b) surface temperature distribution (3D) after 5 s. (Reprinted from Wang, W. et al., 2015, Electrical actuation properties of reduced graphene oxide paper/epoxy-based shape memory composites, *Composites Science and Technology*, 106: 20–24, with permission from Elsevier.)

The internal structure and morphology of multilayered nanopaper embedded in an SMPC at an accelerating voltage of 10 keV is shown in Figure 9.10. The morphology reveals the presence of continuous three-dimensional nodes in the system that form a continuous path for electrical conduction (Figure 9.10a,c,d). The layered arrangement of nanopaper forms an orderly means of heat distribution to the bulk of the SMPC, as evident from Figure 9.10b.

The mechanical properties and performance of the polymer composites are enhanced by the interfacial bonding of the matrix and the multilayered nanopaper. Except for the same problems with agglomeration as seen in the case of CNTs, randomly distributed nanopaper has advantages over other fillers in isotropic heating.

It is interesting to note that as the number of layers increases, so too does the conductivity, and it becomes an asymptote for the axis as the number of

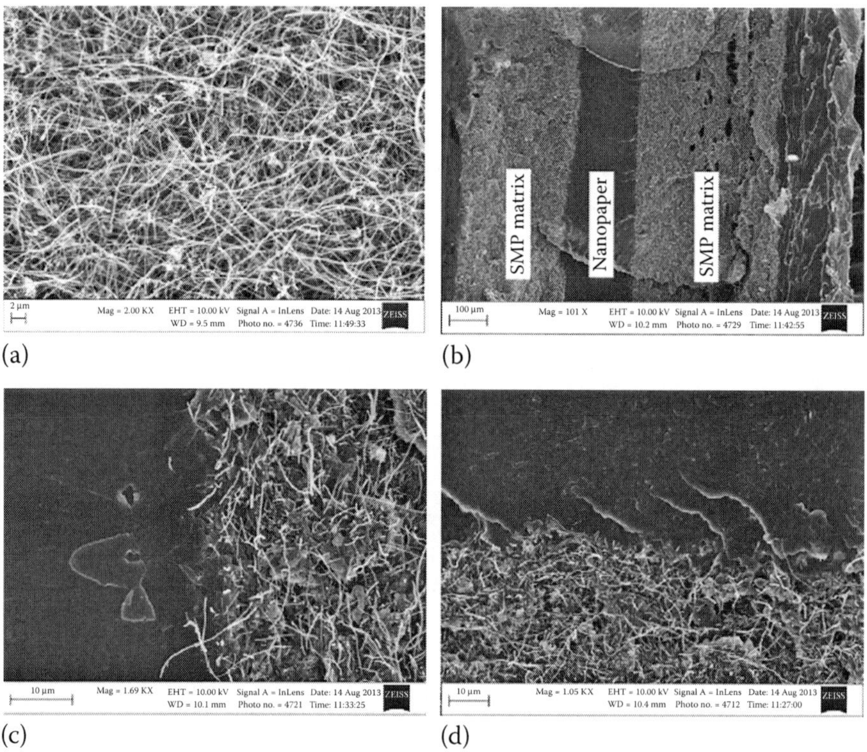

FIGURE 9.10
(a) Morphology of randomly distributed multilayered nanopaper. (b) Structure of multilayered nanopaper-embedded SMPC. (c) and (d) Structure and morphology of the interface between the multilayered nanopaper and the SMP matrix. (Reprinted from Lu, H. et al., 2014, Self-assembled multilayered carbon nanofiber nanopaper for significantly improving electrical actuation of shape memory polymer nanocomposite, *Composites Part B: Engineering*, 59: 191–195, with permission from Elsevier.)

layers reaches four (Figure 9.11a). A further increase in the number of layers could probably bring the curve closer to the horizontal axis in the plot.

As the number of layers of nanopaper increases, the resistance of the system decreases, resulting in increased current as per Ohm's law ($V = IR$). The overall power thus increases ($P = VI$) as the number of orderly layers of nanopaper increases (Figure 9.11b).

In a four-layered nanopaper composite system, the maximum temperature was reached in 80 s (Figure 9.12). The thermal image confirms that the heat transfer in the bulk is uniform and more effective than most of the other types of fillers studied. The T_g of the composite also showed an improvement (narrow range of T_g) as the number of layers increased.

Thus, CNFs, CB, CNTs, and CNP have shown enhancements in electrical and thermal conductivity, heat transfer inside the matrix, and mechanical properties, and improvements in specific strength compared with other dense fillers. As in the literature, the dispersion of fillers in the case of particle-/fiber-reinforced SMPs is the main challenge to be overcome without affecting the other required properties of the material, and the actuation mechanism of this SMPC needs to be discussed in terms of an optimal electric power supply for the system (Yan et al., 2009).

9.2 Actuation Mechanism

As in previous discussions, the incorporation of CNTs/CNFs/CB/CNP into various SMP matrices, or the monolithic synthesis of conductive polymeric material, provides a simplistic alternative to making SMPs electrically

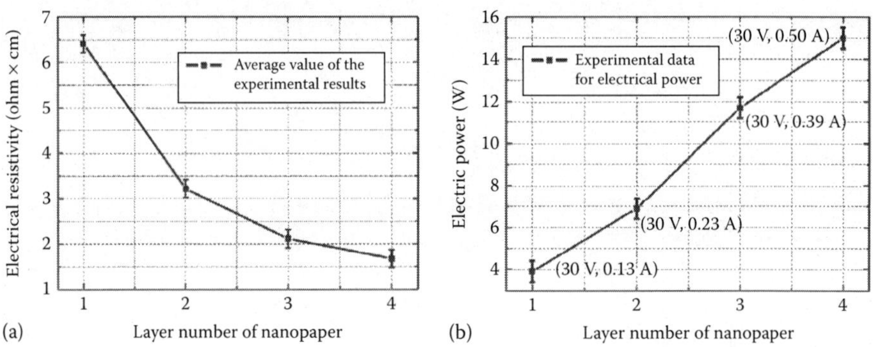

FIGURE 9.11
(a) Electrical resistivity curves of CNF nanopaper. (b) Electric power applied as a function of layers of nanopaper for a constant electric voltage of 30 V. (Reprinted from Lu, H. et al., 2014, Self-assembled multilayered carbon nanofiber nanopaper for significantly improving electrical actuation of shape memory polymer nanocomposite, *Composites Part B: Engineering*, 59: 191–195, with permission from Elsevier.)

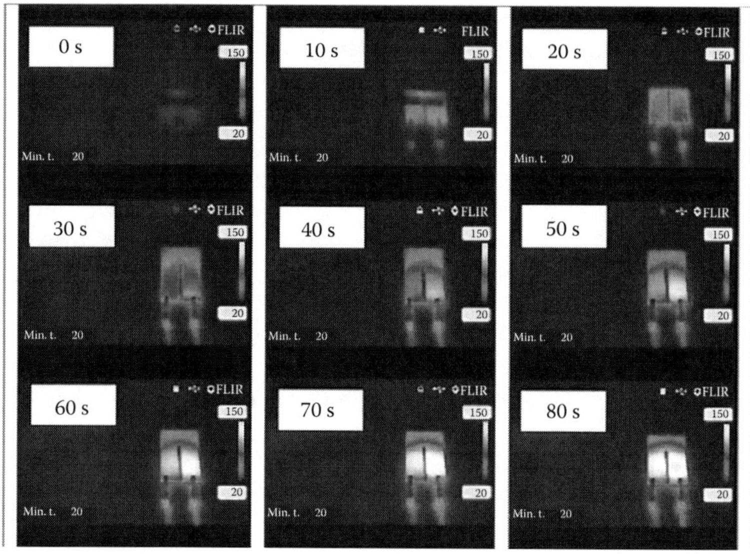

FIGURE 9.12
Thermal imaging of shape recovery and temperature distribution of SMP nanocomposite with four-layered nanopaper. (Reprinted from Lu, H. et al., 2014, Self-assembled multilayered carbon nanofiber nanopaper for significantly improving electrical actuation of shape memory polymer nanocomposite, *Composites Part B: Engineering*, 59: 191–195, with permission from Elsevier.)

sensitive. When the filler content reaches a critical amount, the electrical conductivity is optimized for the composites, and upon excitation with an appropriate electrical field, an electrical current traverses the carbon network, resulting in Joule heating of the internal structure and causing the temperature inside the polymer composite to rise above the T_g. Hence, the SMPC can be deformed to a temporary shape that can be deployed on further heating, when required.

Electric currents can be applied via various methods, such as direct heating, microwave heating, and inductive heating, as per researchers in this field (Pilate et al., 2016). Direct electrical heating is achieved by contact between the surface of the composite and the electrical source through wire. Microwave heating is a noncontact mode of heating that is achieved by the interaction between the molecules and an electromagnetic field with a frequency of 0.3–300 GHz. This is one of the fastest ways to increase the temperature of the material where volumetric heating can be ensured. The filler content plays a vital role in this method of heating as the energy of the microwave is absorbed directly by the fillers. The formation of hotspots on the material due to concentration/overlaps of wave energy is one of the disadvantages of microwave heating. The principle of induction heating is that by passing an alternating current through a magnet, an eddy current is created in the material, thereby resulting in the heating of the bulk.

Material conductivity can enhance the heating rate when this method is employed. Material thickness is a limiting factor, as thicker materials may be heated up nonuniformly due to the uneven distribution of eddy current heating. In induction heating, the power requirement has a direct relationship to the required heat.

Space structures that operate on noncontact shape memory actuation should be programmed with specific stimuli-response temperatures, as unexpected surges of electric charges in space can prematurely trigger the actuation. This requires the SMP's temporary shape training to be specific to a current/field intensity that is different from the operating conditions.

The following chapter summarizes the discussion on shape memory materials and discusses the future prospects and possibilities of further research in the field.

References

Aguilar, J. O., Bautista-Quijano, J. R., and Avilés, F., 2010. Influence of carbon nanotube clustering on the electrical conductivity of polymer composite films, *Express Polymer Letters*, 4(5): 292–299.

Ajayan, P. M. and Tour, J. M., 2007. Nanotube composites, *Nature*, 447(28): 1066–1068.

Deng, Z., Guo, Y., Zhao, X., Li, L., Dong, R., Guo, B., and Ma, P. X., 2016. Stretchable degradable and electroactiveshape memory copolymers with tunable recovery temperature enhance myogenic differentiation, *Acta Biomaterialia*, 46(1): 234–244.

Jiang, Q., Wang, X., Zhu, Y., Hui, D., and Qiu, Y., 2014. Mechanical, electrical and thermal properties of aligned carbon nanotube/polyimide composites, *Composites Part B: Engineering*, 56: 408–412.

Lan, X., Leng, J. S., Liu, Y. J., and Du, S. Y., 2008. Investigate of electrical conductivity of shape-memory polymer filled with carbon black, *Advanced Materials Research*, 47–50: 714–717.

Leng, J., Lan, X., Liu, Y., and Du, S., 2011. Shape-memory polymers and their composites: Stimulus methods and applications, *Progress in Materials Science*, 56(7): 1077–1135.

Lau, K.-T. and Hui, D., 2002. Effectiveness of using carbon nanotubes as nanoreinforcements for advanced composite structures, *Carbon*, 40(9): 1605–1606.

Lu, H., Huang, W. M., Liang, F., and Yu, K., 2013. Nanoscale design of nanosized particles in shape-memory polymer nanocomposites driven by electricity, *Materials*, 6(9): 3742–3754.

Lu, H., Liang, F., Yao, Y., Gou, J., and Hui, D., 2014. Self-assembled multi-layered carbon nanofiber nanopaper for significantly improving electrical actuation of shape memory polymer nanocomposite, *Composites Part B: Engineering*, 59: 191–195.

Luo, X. and Mather, P. T., 2010. Conductive shape memory nanocomposites for high speed electrical actuation, *Soft Matter*, 6(10): 2146–2149.
Meng, Q. and Hu, J., 2009. A review of shape memory polymer composites and blends, *Composites: Part A*, 40: 1661–1672.
Park, J. M., Kim, S. J., Jang, J. H., Wang, Z., Kim, P. G., Yoon, D. J., Kim, J., Hansen, G., and DeVries, K. L., 2008. Actuation of electrochemical, electro-magnetic, and electro active actuators for carbon nanofiber and Ni nanowire reinforced polymer composites, *Composites Part B: Engineering*, 39(7–8): 1161–1169.
Pilate, F., Toncheva, A., Dubois, P., and Raquez, J. M., 2016. Shape-memory polymers for multiple applications in the materials world, *European Polymer Journal*, 80: 268–294.
Qi, X., Dong, P., Liu, Z., Liu, T., and Fu, Q., 2016. Selective localization of multi-walled carbon nanotubes in bi-component biodegradable polyester blend for rapid electro active shape memory performance, *Composites Science and Technology*, 125: 38–46.
Wang, W., Liu, D., Liu, Y., Leng, J., and Bhattacharyya, D., 2015. Electrical actuation properties of reduced graphene oxide paper/epoxy-based shape memory composites, *Composites Science and Technology*, 106: 20–24.
Wang, X. L., Oh, I. K., and Kim, J. B., 2009. Enhanced electromechanical performance of carbon nano-fiber reinforced sulfonated poly(styrene-b-[ethylene/butylene]-b-styrene) actuator, *Composites Science and Technology*, 69(13): 2098–2101.
Yan, Z., Yuexin, D., Lu, Y., and Fengxia, G., 2009. The dispersion of SWCNTs treated by dispersing agents in glass fiber reinforced polymer composites, *Composites Science and Technology*, 69(13): 2115–2118.
Zhang, L., Wan, C., and Zhang, Y., 2009. Morphology and electrical properties of polyamide 6/polypropylene/multi-walled carbon nanotubes composites, *Composites Science and Technology*, 69(13): 2212–2217.

Bibliography

Ahir, S. V., Huang, Y. Y., and Terentjev, E. M., 2008. Polymers with aligned carbon nanotubes: Active composite materials, *Polymer*, 49(18): 3841–3854.
Ajayan, P. M., Schadler, L. S., Giannaris, C., and Rubio, A., 2000. Single-walled carbon nanotube–polymer composites: Strength and weakness, *Advanced Materials Research*, 12(10): 750–753.
Behl, M. and Lendlein, A., 2007. Shape-memory polymers, *Materials Today*, 10(4): 20–28.
Chaterji, S., Kwon, K., and Park, K., 2005. Smart polymeric gels: Redefining the limits of biomedical devices, *Biophysical Chemistry*, 257(5): 2432–2437.
Du, H. Y., Liu, L. W., Chen, F. L., Liu, Y. J., and Leng, J. S., 2014. Design and manufacturing smart mandrels using shape memory polymer, *Proceedings of International Conference on Computational Methods*, 5: 1–8.
Dyana Merline, J. and Reghunadhan Nair, C. P., 2012. Carbon/epoxy resin based elastic memory composites, *Eurasian Chemico-Technological Journal*, 14(3): 227–232.

Fang, G., Peng, F., and Lee, W. 2010. Shape memory polymer composite and its applications in deployable space truss structures, *Materials and Structures Symposium (C2)*, 5: p. 6718.

Gall, K., 2002. Shape memory polymer nanocomposites, *Acta Materialia*, 50(20): 5115–5126.

Icardi, U. and Ferrero, L., 2009. Preliminary study of an adaptive wing with shape memory alloy torsion actuators, *Materials and Design*, 30(10): 4200–4210.

Kuksenok, O., Balazs, A. C., Liu, Y., Harris, V., Nan, H. Q., Mujica, M., Vasquez, Y., et al., 2016. Stimuli-responsive behavior of composites integrating thermo-responsive gels with photo-responsive fibers, *Materials Horizons*, 3(1): 53–62.

Lan, X., Zhang, R., Liu, Y., and Leng, J., 2011. Fiber reinforced shape-memory polymer composite and its application in deployable hinge in space, *Proceedings of 52nd AIAA/ASME/ASCE/AHS/ASC Structures, Structural Dynamics and Materials Conference*, Denver, CO, pp. 1–10.

Leng, J., Lu, H., and Du, S., 2008. Conductive shape memory polymer composite technology and its applications in aerospace, *Proceedings of 49th AIAA/ASME/ASCE/AHS/ASC Structures, Structural Dynamics, and Materials Conference; 16th AIAA/ASME/AHS Adaptive Structures Conference*, Schaumburg, IL, pp. 1–8.

Liu, Y., Du, H., Liu, L., and Leng, J., 2014. Shape memory polymers and their composites in aerospace applications: A review, *Smart Materials and Structures*, 23(2): 023001.

Luo, H., Li, Z., Yi, G., Zu, X., Wang, H., Huang, H., Wang, Y., Liang, Z., and Zhang, S., 2014. Multi-stimuli responsive carbon nanotube-shape memory polymeric composites, *Materials Letters*, 137: 385–388.

Ma, L., Zhao, J., Wang, X., Chen, M., Liang, Y., Wang, Z., Yu, Z., and Hedden, R. C., 2015. Effects of carbon black nanoparticles on two-way reversible shape memory in crosslinked polyethylene, *Polymer (United Kingdom)*, 56: 490–497.

Marconnet, A. M., Yamamoto, N., Panzer, M. A., Wardle, B. L., and Goodson, K. E., 2011. Thermal conduction in aligned carbon nanotube–polymer nanocomposites with high packing density, *American Chemical Society Nano*, 5(6): 4818–4825.

Mohd Jani, J., Leary, M., Subic, A., and Gibson, M. A., 2014. A review of shape memory alloy research, applications and opportunities, *Materials and Design*, 56: 1078–1113.

Qing, Q. N., Ohki, T., Ohsako, N., and Iwamoto, M., 2013. Thermo-mechanical behavior of smart composites with shape memory polymer, *Proceedings of International Conference on Composite Materials: Smart Materials and Structures*, Beijing, China, ID-1332.

Ratna, D. and Karger-Kocsis, J., 2008. Recent advances in shape memory polymers and composites: A review, *Journal of Materials Science*, 43(1): 254–269.

Santhosh Kumar, K. S., Biju, R., and Reghunadhan Nair, C. P., 2013. Progress in shape memory epoxy resins, *Reactive and Functional Polymers*, 73(2): 421–430.

Santo, L., Quadrini, F., Accettura, A., and Villadei, W., 2014. Shape memory composites for self-deployable structures in aerospace applications, *Procedia Engineering*, 88: 42–47.

Shi, Y., Yoonessi, M., and Weiss, R. A., 2013. High temperature shape memory polymers, *Macromolecules*, 46(10): 4160–4167.

Sokolowski, W. M., 2004. US Patent no. US 006702976 B2, Cold hibernated elastic memory self-deployable and rigidizable structure and method therefor, March 9.

Sokolowski, W. M. and Hayashi, S., 2003. Applications of cold hibernated elastic memory (CHEM) structures, *Proceedings of SPIE: The International Society for Optical Engineering*: 5056: 534–544.

Tong, T. H., 2007. US Patent no. US 007276195 B1: Maleimide based high temperature shape memory polymers, October 2.

Vernon, L. B. and Vernon, H. M., 1941. US Patent no. US 2234993: Producing molded articles such as dentures from thermoplastic synthetic resins, Mar 18.

Wang, X. L., Oh, I. K., and Kim, J. B., 2009. Enhanced electromechanical performance of carbon nanofiber reinforced sulfonated poly(styrene-b-[ethylene/butylene]-b-styrene) actuator, *Composites Science and Technology*, 69(13): 2098–2101.

Wei, Z. G., Tang, C. Y., and Lee, W. B., 1997. Design and fabrication of intelligent composites based on shape memory alloys, *Journal of Materials Processing Technology*, 69 (1–3): 68–74.

Yang, D., 2000. Shape memory alloy and smart hybrid composites: Advanced materials for the 21st century, *Materials and Design*, 21(6): 503–505.

Yin, W., Fu, T., Jingcang, L., and Leng, J., 2009. Structural shape sensing for variable camber wing using FBG sensors, *Proceedings of SPIE: The International Society for Optical Engineering*, 7292: 1–11.

Zhang, F., Zhang, Z., Zhou, T., Liu, Y., and Leng, J., 2015. Shape memory polymer nanofibers and their composites: Electrospinning, structure, performance, and applications, *Frontiers in Materials*, 2: 1–10.

10

Discussions and Future Prospects

This chapter summarizes and concludes the discussions from the previous chapters of the book and explores the future prospects in the field of shape memory smart materials for application-specific material development.

10.1 Smart Materials

Technological advancement is possible only through the development of novel materials to form smart structures and systems for specific applications. The evolution of such smart materials has its origin in the unending quest to harness the efficiency of the mechanical systems available in nature. As discussed, smart materials sense stimuli, process information, and actuate a response in various ways.

One of the most novel and efficient methods of achieving smart actuation is through stimuli-responsive shape memory materials (SMMs), which could replace many complex mechanisms, as discussed in previous chapters. Shape memory has progressed through alloys, ceramics, and gels to reach polymers, and is still under a continuous process of evolution, irrespective of the end application.

10.2 Shape Memory Materials

As previously discussed, the field of smart materials is interdisciplinary, involving material science, metallurgy, polymer chemistry, physics, and mathematics. This book has explored the evolution of SMMs, taking in the concepts of shape memory alloys, ceramics, gels, and polymers, the intrinsic and induced mechanisms that result in smart behavior, their characterizations and advantages over existing systems, and their applications.

Being a novel branch of material science, more experiments and mathematical backup are required to make better predictions and to understand and apply the shape memory effect (SME) in various materials.

10.3 Shape Memory Polymer Composites

SMMs with various properties are combined together to make shape memory composites. Polymers, generally insulating materials by nature, are made to conduct/respond to temperature/magnetism/electricity by including conducting fillers or material components. These components can be a conductive filler material or a conductive polymer, as per the published literature (Aguilar et al., 2010; Ajayan et al., 2000; Dyana Merline and Reghunadhan Nair, 2012; Lan et al., 2008).

Composites formed from SMPs are the solutions to many applications that require lightweight, high-precision actuations by remote access. Nanomaterials such as carbon black (CB), carbon nanofibers (CNFs), carbon nanotubes (CNTs), and so on are some of the novel additions to fillers in such composites, resulting in nanocomposites, as discussed in previous chapters of this book.

The creation and optimization of nanocomposites is a huge challenge to be addressed by the scientific community, as they have huge potential in future technologies. Nanocomposites or carbon nanopolymer composites, with application-specific properties and multifunctional attributes such as self-healing, self-strengthening, self-cleansing, remote stimulation, and so on, have desirable features for aerospace applications.

The interfaces between matrices and fillers are an area of significant interest, and the development of characterization methods to understand the atomic-scale structure of these interfaces is required to enhance the current advantages. Thus, a better understanding of these interfaces through detailed experimental, analytical, and modeling studies is essential for the efficient usage of these material systems.

Complex composite nanoscale building blocks such as nanoparticles, nanofibers, nanotubes, and nanopapers that have evolved out of graphene or nanotube stacks are the latest trend in the field of shape memory nanomaterials. The scientific community has invested its resources in carefully learning about the intricate behaviors of nanofillers to improve their thermal, electrical, and strength parameters, their shape recovery, and so on with various matrix materials. Parallel to the exploration of material optimization, cost is one of the critical aspects that dictates application-based manufacturing. The availability of high-efficiency materials for experiments provides a better platform from which to seek out solutions for many unattended challenges in the field of SMMs.

Studies on shape memory polymers (SMPs) with nanomaterial fillers can include multiple fillers in varying proportions to optimize and categorize application-based materials. Understanding a specific property can enable designers/material scientists to form a catalogue of shape memory polymer composites (SMPCs) that can be used directly off the shelf. Chemical reactions can even be initiated between different nanotube types/nanomaterials within the same matrix, resulting in newer thermal or electrical properties.

The tuning of switching temperature is an area that has large scope for further research in the design of new SMPs/SMPCs. Polymerization time, cross-link density, the addition sequence of monomers and fillers, surfactants/scaffolds, temperature, stirring speed, and so on play vital roles in deciding the final polymer. Varying these parameters results in the final product having different properties. Surface-modifying agents for fillers is another area of study that is open for exploration, since the modification of filler surfaces affects shape memory behavior.

The high-temperature applications of SMPCs are still at the laboratory scale and, therefore, are a significant research area. High switching temperatures combined with excellent thermal stability and good mechanical properties are prerequisites for SMPCs in aerospace applications. The variety of triggering stimuli, such as inductive heating, Joule heating, infrared irradiation, photothermally induced heating, and so on, improves the choice of SMMs for specific applications.

Processing difficulty is one of the drawbacks of SMPs, as most of the polymer structures described are ideal candidates for laboratory studies, whereas large-scale production or commercialization is difficult. The advent of 3D and even higher-dimension printing techniques (e.g., 4D printing, the shape morphing of 3D printed structures, is explained in Section 5.5) could benefit from improving the material properties of SMPs.

Improvements in the constrained recovery efficiency of SMPs is one of the less explored fields. The development of thin-film SMMs with electrical or thermal actuation requires the shape memory film to pull its own weight to deform and recover, which needs further study.

Control over the spatial distribution of SMPs allows for the fabrication of complex architectures in SMPCs. The possibility of fibers grown into fabrics has attracted the attention of the textile industry and could be an extension of additive manufacturing. Smart fabrics and smart clothing are areas of research pertaining to flexible SMPC studies and are one of the demanding research fields in current smart/intelligent world scenarios. Optimizing the influence of switching temperature on the overall energy balance of SMPs will open up wider possibilities in aerospace applications.

Multiple stimuli–responsive SMMs are novel additions to the range of SMMs, where the polymer can give different responses to different triggering stimuli (light, heat, moisture, magnetism, etc.). Surface-modified superparamagnetic nanoparticles in SMP matrices have been found to facilitate electromagnetically triggered SME. Multiple (two-way, three-way, etc.) SME is a recent innovation in which the material recovers its original shape through more than one intermediate, temporary shape.

Scientists have shown that chemically cross-linked thermoset SMPs are better candidates for realistic applications than physically cross-linked thermoplastics, due to their excellent shape fixity and recovery ratios, higher transition temperatures, and better chemical and thermal stability (Liu et al., 2007). A combination of the advantages of thermosets and thermoplastics in shape memory is a trending research area in the field of multi-ink printing in additive manufacturing.

As shown in many research publications, one of the major issues pertaining to SMPCs is their response/recovery time (Leng et al., 2011; Gall, 2002). Slow response to stimuli is one of the limitations being addressed by researchers across the globe. Hence, the improvement of recovery speed is a major area of interest. It has been found that repeated shape memory cycles reduce the tensile strength of materials (about 6% strength reduction for epoxy/carbon fiber systems). Thus, fatigue strength is another area of SMPC research that needs further exploration. A correlation between the degree of cure and the shape memory property of polymer composites would help to classify materials for specific applications to help designers choose a particular material for a particular purpose. A few research works have reported the effect of cure degree on the shape memory characteristics of epoxy systems and observed an enhanced T_g (Santhosh Kumar et al., 2013). This was achieved with the right type and amount of curing agent (increasing the curing agent content increases SME), as the cross-linking density is directly influenced by the curing compound.

Since almost all studies on CNTs (MWNTs/SWNTs) end up with the dispersion of nanoparticles in the matrix, the various means of effectively dispersing CNTs into matrices will be the key to future research works.

Apart from the conventional approaches like the addition of CNFs, CB, and CNTs (as discussed in the previous chapter), the multiresponsiveness of SMPCs can be enhanced by incorporating non-carbon-based nanosized fillers (noble metal–based nanostructures, metal oxides, and cellulose nanocrystals). Shape memory hybrids with optimal alloy–ceramic–gel–polymer combinations can offer the properties of each component, resulting in a high-grade smart material.

10.4 Electroactive SMPs

Electrical stimuli–responsive SMPs could be the future, as they can bring about the precise remote actuation of temperature-sensitive systems. The conductivity of the matrix or the inclusion of conductive parts in the composite facilitates electroactivity. Inaccessible operations/triggers/actuations in various engineering disciplines can reap the benefits of electroactive SMPs (EASMPs).

Advanced design concepts for the integration of textile electronics in composites with electrically actuated intelligent systems have been studied for use in the textile industry (Varga and Tröster, 2014). Stacked layers of electronically embedded textiles, printed circuits on flexible substrates, nodal point control units, and so on could result in e-composites with sensing

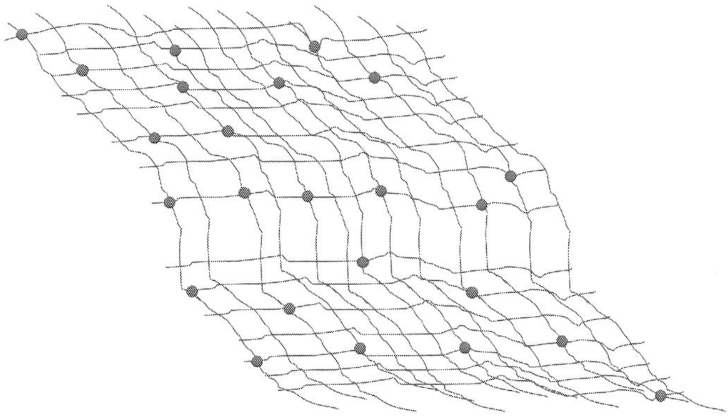

FIGURE 10.1
Conceptual representation of textile composite with nodal electronic unit blocks.

abilities (Figure 10.1). Electrical actuation, which is thought to be more advantageous than direct heating and other stimuli, could be the future of the remote access/triggering of shape memory systems used in aerospace or other high-precision, inaccessible areas.

The applications of EASMPs are not limited to microsensors and microactuators but could provide cheaper, more accurate, and better alternatives to devices already in use. Electroactive biodegradable shape memory copolymers with tunable recovery temperatures and superstretchable properties are being explored for biomedical applications in the area of muscle tissue engineering. Excellent strain at break, electroactivity, and good shape memory properties with suitable recovery temperatures around body temperature, high recovery ratios, good fixity ratios, and short recovery times are the expected properties of such SMPCs. Similar to the natural responses of human body parts to the electrical signals from the brain, artificial elements that are sensitive to small electrical actuation stimuli shall serve as supplements to human body parts such as muscles and tissues with elastic memory.

Establishing a standard regarding the electrical input signals and mechanical output will help designers choose materials for a specific use.

10.5 4D Printing

Three-dimensional printing is the trending subject in manufacturing as it is gaining popularity in almost all scientific disciplines dealing with product realization. The concept is the extrusion of molten polymers that upon cooling and solidifying form a 3D structure. The use of additive manufacturing enables the production of complex geometries that are difficult to produce with other techniques.

A 3D-printed sheet of an octahedron (Figure 10.2a) that evolves into a self-folding configuration on hydroactivation has been demonstrated by Skylar Tibbits et al. at MIT. This concept has been called *4D printing* as the fourth dimension is defined by the evolution in time from a 3D-printed shape. On contact with water, this hygroscopic 2D shape turns into a 3D configuration of an octahedron by self-folding (Figure 10.2b,c).

Thus, researchers and scientists have realized 3D structures with dynamic behavior using multimaterial inkjet printers (Tibbits, 2014). Four-dimensional printing concerns the material that evolves after 3D printing. Soft robotics and flexible electronics demand the fabrication of SMPs in any geometry that exhibit shape memory behavior for use in actuations. It has been proven that any cross-linkable polymer with the property of shape memory and controllable melt viscosity can be printed as a 3D structure.

Future reconnaissance on the possibilities of fabricating complex shape memory structures with a viscous melt using a 3D printer will provide the option of ready-to-use actuators.

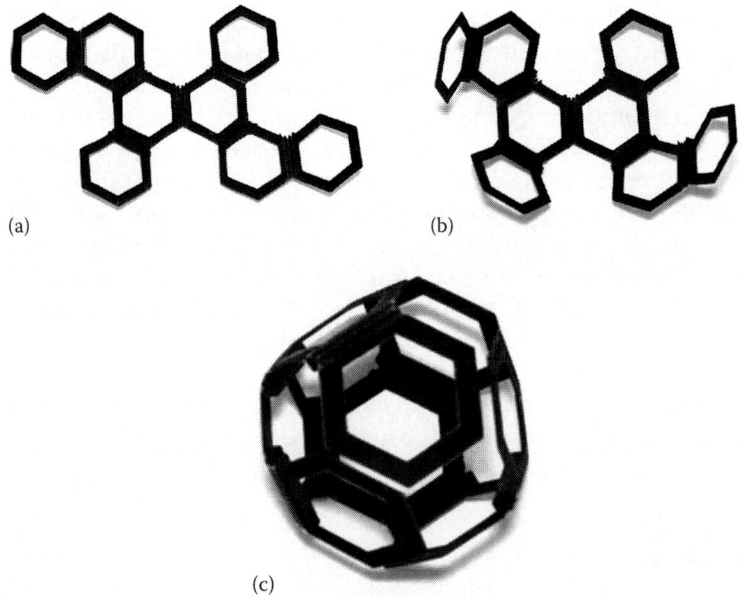

FIGURE 10.2
Self-folding hydroactive shape memory truncated octahedron. (a) Open configuration. (b) Hydroactivated to fold. (c) Folded configuration in water. (Reprinted from Tibbits, S., 2014, 4D printing: Multi-material shape change, *High Definition: Zero Tolerance in Design and Production,* 84(1): 116–121, with permission from John Wiley & Sons.)

References

Aguilar, J. O., Bautista-Quijano, J. R., and Avilés, F., 2010. Influence of carbon nanotube clustering on the electrical conductivity of polymer composite films, *Express Polymer Letters*, 4(5): 292–299.

Ajayan, P. M., Schadler, L. S., Giannaris, C., and Rubio, A., 2000. Single-walled carbon nanotube–polymer composites: Strength and weakness, *Advanced Materials Research*, 12(10): 750–753.

Dyana Merline, J. and Reghunadhan Nair, C. P., 2012. Carbon/epoxy resin based elastic memory composites, *Eurasian Chemico-Technological Journal*, 14(3): 227–232.

Gall, K., 2002. Shape memory polymer nanocomposites, *Acta Materialia*, 50(20): 5115–5126.

Lan, X., Leng, J. S., Liu, Y. J., and Du, S. Y., 2008. Investigate of electrical conductivity of shape-memory polymer filled with carbon black, *Advanced Materials Research*, 47–50: 714–717.

Leng, J., Lan, X., Liu, Y., and Du, S., 2011. Shape-memory polymers and their composites: Stimulus methods and applications, *Progress in Materials Science*, 56(7): 1077–1135.

Liu, C., Qin, H., and Mather, P. T., 2007. Review of progress in shape memory polymer, *Journal of Material Chemistry*, 17(16): 1543–1558.

Santhosh Kumar, K. S., Biju, R., and Reghunathan Nair, C. P., 2013. Progress in shape memory epoxy resins, *Reactive and Functional Polymers*, 73(2):421–430.

Tibbits, S., 2014. 4D printing: Multi-material shape change, *High Definition: Zero Tolerance in Design and Production*, 84(1): 116–121.

Varga, M. and Tröster, G., 2014. Designing an interface between the textile and electronics using e-textile composites, *Proceedings of the 2014 ACM International Symposium on Wearable Computers Adjunct Program (ISWC '14)*, Seattle, WA, pp. 255–260.

Bibliography

Ajayan, P. M. and Tour, J. M, 2007. Nanotube composites, *Nature*, 447(28): 1066–1068.

Deng, Z., Guo, Y., Zhao, X., Li, L., Dong, R., Guo, B., and Ma, P. X., 2016. Stretchable degradable and electroactive shape memory copolymers with tunable recovery temperature enhance myogenic differentiation, *Acta Biomaterialia*, 46(1): 234–244.

Huang, W. M., Yang, B., and Qing, F. Y., 2017. *Polyurethane Shape Memory Polymers*, CRC Press, Boca Raton, FL.

Javid, S. and Vatankhah, M., 2016. Fundamental investigation on the mechanisms of shape memory polymer reversibility, *International Journal of Chemical Studies*, 4(4): 43–45.

Jiang, Q., Wang, X., Zhu, Y., Hui, D., Qiu, Y., 2014. Mechanical, electrical and thermal properties of aligned carbon nanotube/polyimide composites, *Composites: Part B*, 56: 408–412.

Liu, Y., Wang, X., Lan, X., Lv, H., and Leng, J., 2008. Shape memory polymer composite and its application in deployable hinge for space structure, *Proceedings of SPIE: Sensors and Smart Structures Technologies for Civil, Mechanical, and Aerospace Systems*, 6932(2): 10–18.

Merline, J. D., Reghunadhan Nair, C. P., and Ninan, K. N., 2008. Synthesis, characterization, curing and shape memory properties of epoxy-polyether system, *Journal of Macromolecular Science, Part A: Pure and Applied Chemistry*, 45(4): 312–322.

Michaud, V., 2004. Can shape memory alloy composites be smart? *Scripta Materialia*, 50(2): 249–253.

Van Humbeeck, J., 1999. Non-medical applications of shape memory alloys, *Materials Science and Engineering: A*, 273–275: 134–148.

Varadan, V. K., Vinoy, K. J., and Gopalakrishnan, S., 2006. *Smart Material Systems and MEMS: Design and Development Methodologies*, Wiley, Chichester, UK.

Index